Biorhythms at Your Fingertips

BIORHYTHMS AT YOUR FINGERTIPS

James Roche

JAVELIN BOOKS

POOLE · DORSET

First published in the UK 1985 by Javelin Books,
Link House, West Street,
Poole, Dorset, BH15 1LL

Copyright © James Roche 1985

Distributed in the United States by
Sterling Publishing Co., Inc.,
2 Park Avenue, New York, NY 10016

British Library Cataloguing in Publication Data

Roche, James
  Biorhythms at your fingertips.
  1. Biological rhythms
  I. Title
  612'.022   BF637.B55

ISBN 0 7137 1562 6

Typeset by Poole Typesetting (Wessex) Ltd, Bournemouth

Printed in England by Guernsey Press

CONTENTS

Dedicated to Cynthia, for all her encouragement.

ACKNOWLEDGEMENTS

My thanks to the following publishers for permission to quote extracts: George Allen & Unwin (Publishers) Ltd for 'How Not To Kill Yourself'; Viking Penguin Inc. for 'Mr Thompkins Inside Himself'; The Longman Group Ltd for *Roget's Thesaurus;* Macmillan London Ltd for 'Fabric Of The Universe'; and Maurice Temple Smith Ltd for 'Body Time'. My thanks also to Garnstone Press Ltd for 'Biofeedback', whom I have not been able to contact.

INTRODUCTION

Some get their *kicks* from *jogging,* others from *yoga* or *TM.* If they go to a *shrink* with their *hang-ups,* he will set them *role-playing* in an *encounter group,* or is this just an *ego trip?* If they consult a *naturopath,* a *barefoot doctor* or a *paramedic* with a *bleeper,* he or she may study their *biorhythms,* and prescribe *hormone therapy, pep pills* or *EST.* If they become unconscious, they may be put on a *life support system* or given the *kiss of life.*

ROGET'S THESAURUS – Susan Lloyd

Just to put the record straight, I'm not a naturopath, barefoot doctor, or a paramedic with a bleeper! But, with this book you will be able to study your own biorhythms without embarking on the needless expense of consulting an expert: you will quite easily become your own expert.

Biorhythm first started with the investigation of mental and physical health in hospital patients and also academic performance in school and college students. Now its principles are widely known throughout Japan, Germany, Switzerland, Britain, and America. From choosing an advantageous date for major surgery to relaxation programmes for stress, and coaching schedules for athletes, its guidelines inspire all of us to increase our self-understanding and improve our relationships with those around us. From its inception nearly ninety years ago, it has become recognized for its contribution to human awareness and potential, although to some – more conventional thinkers – its approach to mankind's rhythmicity is still a little controversial. Nevertheless, the detailed analysis of biorhythms produces results that are of great interest to many people.

Biorhythm has far reaching applications, not only for the individual, but also for closely-woven groups, and of course society at large. Because of the tremendous upsurge in interest towards this important science, there are many who have employed, and gained benefit from, its policies. Popular magazines offer a very reasonable service of biorhythm printouts for a small fee. All that is required is your name, address and date of birth. It is only with the aid of recent computer technology that the demands for this type of service have been met.

'What, then, are these biorhythms like? Do men possess the emotional rhythm? How do these cycles affect us? When do they commence and finish? What can biorhythm do for *me?*' All of these are commonplace enquiries that a newcomer to biorhythm will raise. Throughout this book I hope to deal with each of these fundamental issues, so I shall start, in brief, with the character of these cycles and when they originate. Let's begin at the beginning, which is, for all of us, birth.

The moment that you were born, your body underwent considerable changes: for the first time you had to learn to breathe in oxygen from the air outside your body rather than obtain it from your mother's circulating blood. You might have been coaxed into breathing by being hung upside down, bat fashion, and smacked. (Surely enough to make the most stoic neonate swing into a lung-bursting protest after the warmth and security of being inside the mother.) For the first time your body had to go it alone, with no mother to look after all your needs of nutrition and excretion.

As a result of this enormous stimulation, and transformation of environment, biorhythm theory states that the three biorhythm cycles start at the moment of birth. These three cycles are: the physical cycle, the emotional cycle, and the intellectual cycle. They affect us in diverse ways throughout our lives – as their names suggest – and they also have characteristic times to complete one full rhythm. The physical cycle will complete itself every twenty-three days; the emotional (or sensitivity) cycle will complete itself every twenty-eight days; and finally, the intellectual cycle will complete itself every thirty-three days. It is because of these three differing phase lengths that there are a seemingly infinite number of variations in our day-to-day biorhythmic make-up. Only after fifty-eight years and sixty-seven days will all the combinations of highs, lows, and critical days in these rhythms be exhausted. At that point the cycle then repeats itself and exhibits wave patterns that are evident from the time of birth.

This approximately fifty-eight year period is called the *biorhythmic span*, and during this time each of us will experience highs, lows, critical days, double critical days, and last but by no means least, the 'dreaded' triple critical day. (For those of you who know me, my next triple critical day is on July 7, 1986, so be nice to me!) This inauspicious event occurs less than once a year – for that I think most students of biorhythm are eternally grateful! However, with careful planning and a dose of common sense, even these irksome days can be negotiated with positive results.

Considering the rhythms themselves, all the cycles show the same undulating pattern, although, as indicated, they require varying amounts of time to complete one full cycle. The easiest way to think of these invisible rhythms of our own bodies is to visualize the waves of the sea, each with their own peaks and troughs, following a smooth rhythmic, continuous motion. For those of you who are more mathematically minded, biorhythm theory instructs us that these biological rhythms follow a sine wave relationship. The following illustration gives an example of biorhythm conditions for any baby at birth. The only possible variation being that the whole pattern will be shifted to the left or right according to the date of birth. All three cycles start at this point in an individual's life. This chart depicts a birth time of about midday on the fifth of the month. The physical cycle is represented by a continuous line, the emotional cycle by a dashed line, and the intellectual cycle by a dotted and dashed line. This is the international standard. A natal biorhythm chart is illustrated below.

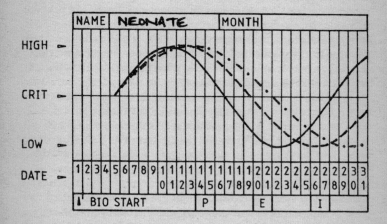

As illustrated, all three cycles will start to rise from the critical line on the day of birth into the positive biorhythm condition. (The day of birth is a triple critical day.) After the peak in each wave has been reached, a decline starts to set in, with the result that each rhythm will pass through the critical line at exactly halfway through its cycle. For the remaining half of the phase, the rhythm will be in a negative biorhythm condition. When the lowest point in the cycle is experienced, the wave form will start to rise again so that the critical line is crossed once more. This is the pattern for one complete cycle and is identical for all three biorhythms. The time taken from start to finish will depend upon the biorhythm concerned.

9

After the first month of life, the waves begin to 'separate' from each other by virtue of their differing frequencies, and thus a much more random pattern is produced. The resultant combination of highs, lows, and critical days are predictably numerous. It is for this reason that these cyclical phenomena in man had to be discovered by a scientific, rational analysis of large numbers of people before the true nature of these inner systems could be evaluated.

BIORHYTHM AIDS

Over the past few years, the natural growth of this science has stimulated the demand for more consumer-orientated products, rather than devices that are suitable for expert use only. Many of these aids have been successfully designed and marketed: some of them more sophisticated than others. They range from simple scales that illustrate biorhythm alignments to high-powered computers that will only handle these specific calculations. If the ardent follower of biorhythm finds the daily reference to paper charts too cumbersome, a Swiss enterprise, Certina, will solve the problem. They have ingeniously managed to incorporate three biorhythm scales (in their internationally accepted colouration) into the face of a wristwatch called the Biostar. Nothing could be simpler, or more élite, for the dedicated supporter. On purchase, the three biorhythm bands are set for the birth date of the owner. Future wave positions are automatically indicated for successive dates as they arrive. A quick glance at this purpose-built watch will graphically reveal highs, lows, and critical days. The positive phases in each cycle are depicted in their standard colours (physical – blue, emotional – red, intellectual – green). The negative phases are left a silver colour. By creating this contrast between silver and each band colouration, critical day crossing points can be easily recognized by a clear dividing line. The only drawback with this biorhythm aid is that it is personal – it can only depict biorhythm conditions for one person, the owner. So, if this particular aspect of the Biostar doesn't appeal to you, as you would also like to study the biorhythms of your friends, Casio will provide you with a calculator called the Biolator. This sophisticated elelctronic calculator will compute anyone's bioprofile for the day of interest, as well as perform the more usual arithmetical routines. Although there is no memory facility for birth data in the Biolator, this is compensated for by an extremely rapid evaluation of an individual's daily changing biorhythm status. You need wait only a second, or two at the most, for the result.

In Japan, an equally rapid record of biorhythm conditions can be obtained by the use of the Biocom 200. This computer determines an individual's biorhythms in seconds and then prints out a 'hard copy' in the form of a small pocket-sized card showing conditions for one month. This particular innovation has proved extremely popular, with its subsequent utilization within insurance companies, traffic police departments, and large transportation firms like Seino Transport and Japan Express. Literally thousands of these cards have been issued to Japanese citizens. Ingenious methods have been employed to bring biorhythm to the people. Taxi drivers display a coloured paper crane in their window (the feathered variety, a symbol of good luck) and bikers show a flag of the correct colouration. Both systems warn other road users of the owner's biorhythm condition. There is hardly an area in Japanese society where biorhythm has not left its characteristic mark. Car accident rates have plummeted despite increased mileage per annum on the highways. Improved road safety has encouraged biorhythmically-based safety training in factories. Life insurance companies have welcomed this science for the obvious benefits that it brings: increased human longevity and decreased accident claims.

In the most recent edition of *Roget's Thesaurus,* Susan Lloyd illustrates the incorporation of new words into our language by means of an amusing construction of situations using these linguistic arrivals. (This is quoted at the head of this chapter.) Words that are capable of transmitting thoughts or ideas that were previously uncalled for. *Biorhythm* now appears in up-to-date dictionaries and is a contraction of *bio*logical and *rhythm.* The word has come into being through the necessity to distinguish this science from the study of other biological cycles, and because of the amount of research into this discovery. Since Fleiss's detection of these biological cycles, many lives have been greatly influenced and have taken on a new, purposeful direction. One of those notable persons, who contributed much to the cause of biorhythm, was a Swede, George Thommen, and it was many years ago that my interest in this discipline was kindled through reading his book, *Is This Your Day?*

FIVE DAYS TO LIVE

An unforgettable event in the relevance of biorhythm occurred as a result of a radio interview in the USA with George Thommen on 11 November, 1960. At that time the actor Clark Gable was ill in hospital recovering from a heart attack that had struck whilst filming with Marilyn Monroe. Thommen had carefully studied the biorhythm trends that were evident for Gable during this month. In particular, Thommen was convinced that his physical rhythm should be observed, and the necessary precautions taken. He reasoned that as Gable's health was already well below par, any critical days in this rhythm would counsel extreme caution. In fact, his speculations were no longer speculations, his 'theory' had already been substantiated: only six days before the interview, Clark Gable had suffered a near-fatal heart attack – experiencing a physically critical day! In view of the actor's weakened plight, and the fact that the physical rhythm has the most frequent occurrence of critical days, there was little time for his strength to build up again in time for the next unstable period: 16 November. Thommen demonstrated and explained on the programme why every precaution against another relapse should be taken – this time it could be Clark Gable's last critical day; all possible efforts should be taken to ensure that 16 November did not pass with the sad news of this actor's death. Unfortunately, the medical profession did not respond to Thommen's request. Perhaps they did not hear; perhaps they did not want to hear. The result was that the necessary equipment for reviving Clark's heart was not immediately available. The consequences everybody is aware of: Clark Gable died on 16 November, 1960 – the very day that Thommen had cautioned.

THOMMEN'S INFLUENCE

Throughout his book, Thommen discusses the theory and practice of biorhythm with equal conviction. To many adherents this book is considered a classic. Its effects were certainly far reaching in the scientific community in the USA, but Thommen's message soon took on an international flavour with his despatch of a copy to the distant shores of Japan in 1965. *Is This Your Day?* so impressed Kitchinosuke Tatai, a professor of physiology at the Institute of Public Health in Tokyo, that he immediately set to work investigating these biological rhythms. This type of research was quite familiar to Tatai, for in 1947 he submitted his doctoral thesis to the University of Tokyo on the human ability to adapt to seasonal changes. Most of Tatai's efforts were channelled into the examination of time-based phenomena in man that were of a biological origin. After nearly twenty years of further research into stress, circadian rhythms, and micro-circulatory rhythms, he ventured into biorhythm, stimulated by Thommen's insight and discoveries into this fascinating human periodicity. Since those early days, Tatai has been one of the main proponents of biorhythm in the Far East, promoting and establishing this discipline for the benefit of his fellow citizens.

Like Kitchinosuke Tatai, *Is This Your Day?* had a profound effect on my perception of human life. I soon found myself examining and applying the theories to my own existence. So as not to become biased in favour towards biorhythm I made a point of keeping a diary of my moods, state of health, and mental energy without knowing my status. I decided that the best line of approach was retrospective analysis: the analysis of past conditions in the light of biorhythm theory. I recorded any information that might subsequently reveal trends and patterns in my thinking, feelings, and physical well-being. At the beginning of each month I analysed my performance over the previous month. Some might argue that even this frequency of retrospection would leave an impression of future conditions for at least the first week in the month and so 'colour' a quarter of my observations. However, it was not long before I started charting friends' biorhythms as well. With this added dimension of working on several charts at a time, I found it nigh impossible to recall my own rhythmic phasing. This I think was a distinct advantage in maintaining an unbiased approach to my own self-assessment.

Over the months, certain trends came to light that corresponded with the principles of biorhythm; the few exceptions being well within any latitude given for personality allowances. For example, some of the rhythms had a greater effect on me than others: the emotional rhythm affected me the least. When interpretating my observations I have to remember my tendency towards applying the majority of my energies to the mental rather than the physical realm. Considering all the attributes that come within the boundaries of the emotional cycle I also realized that although being creative produced a sense of fulfilment, I did not suffer from a distinct vacillation of moods that sometimes characterizes those of a more artistic temperament. Thus, I would be surprised to encounter mood changes that paralleled the emotional cycle. In retrospect, I discovered that during intellectually positive phases, my computer programming was generated faster, tended to have less errors of logic, and my typing was accordingly more accurate. When experiencing intellectually negative phases, I found it much easier to review my work or pursue any desired reading, rather than to tackle any unresolved problems (especially those of a complex, demanding nature). These responses during high and low phases tie in well with what one can expect from this thirty-three day cycle. Considering my creative drive, this was more or less constant from day to day, but I did experience more inspirational 'flashes' during the emotional high periods, so I had a tendency to become more productive at these times. As already mentioned, I did not uncover any noticeable changes of mood or feeling that shadowed this biorhythm – but this was to be expected in view of my temperament. Finally, the physical rhythm did govern the amount of body energy that I felt comfortable in expending. The highs were typified by periods when I became more energetic, and the lows were recognizable as spells of decreased activity. The physically critical days had a pronounced effect on me, especially in the mornings: my responses were definitely sluggish. So, all in all, I found that my behaviour followed biorhythm energy trends quite closely, the few 'anomalies' being quite understandable when viewed in perspective against the background of my personality and lifestyle.

By now you may be thinking, 'That's fine, studying past events and comparing them with critical days, but how can I *improve* my life through the application of these three cycles? Where do I go from here?' Admittedly, retrospective analysis is a captivating pastime and there's no limit to the amount of information that you can glean from your own life or from your friends' lives. That is how biorhythm was born: through inspecting recorded behaviour patterns for a vast number of people. But, all of us want to go forward. Forward by means of self-improvement so that our lives may become richer in experience and thus more fulfilling. So, what's the next step? Like all paths to self-improvement, we have to learn about the system being explained, and without its relevant history, biorhythm would be incomplete.

1 THE BIRTH AND GROWTH OF BIORHYTHM

The simple idea around which science is built is that nature is objective, that it can be comprehended and that the only route to this comprehension is the systematic confrontation of logic and experience. The scientific method is undoubtedly a unique and powerful source of truth, a profound guide (though it has barely begun the work of fusing the inner mental and outer material worlds) to what is.
Fabric of The Universe – Denis Postle

The discovery of biorhythm can be attributed to three men at about the beginning of this century. Each of them were working independently of each other, and yet they all came to the same conclusions: the human body does portray a definite rhythmical fluctuation of moods, physical ability, and intellectual efficiency. It was through these three, scientific, enquiring minds that the cyclical nature of our own bodies was discovered, and through their initial researches the science of biorhythm was born.

WILHELM FLEISS

Dr Wilhelm Fleiss (1859-1928), was an eminent nose and throat specialist at the University of Berlin who later became the chairman of the German Academy of Sciences. Fleiss was a contemporary of the famous Dr Sigmund Freud (1856-1939), who was researching at the University of Vienna where similar discoveries of biorhythm by another pioneer, Hermann Swoboda, were to be made. Freud and Fleiss had many discussions about cyclical variations in mood states in their patients; in fact their discourses were so frequent that in the space of fifteen years they were known to have written just over 180 letters to each other.

Fleiss's theories about the physical and emotional cycles were partly based on his investigations into the human cell. By looking at this microscopic unit of life, Fleiss was able to draw conclusions about the nature of our sexuality. He reasoned that the masculine cycle of twenty-three days and the feminine cycle of twenty-eight days were present in each individual (whether male or female) because of the duality of our sexual nature, even at the cellular level. He concluded that within each human being there was all the necessary information for both female and male characteristics. One of his illustrations of this was that the male still retains nipples although they serve no glandular function as in the female.

At the time of Fleiss's ideas about human bisexuality, the study of hormones was not advanced – for that reason many scholars of his time found his theories a little too far fetched to be believed. There was no evidence in other fields of medicine that his speculations might be valid. However, present-day investigations reveal that both men and women secrete male and female hormones, although the balance between these two 'opposing' groups of biologically active substances is quite different in each sex. In men, there is up to twice the amount of male hormones released into the blood system as there are female hormones. In women, the difference is more obvious – there is about five times the quantity of female hormones present in the body as there are male hormones. Of our recent findings, these reveal that although many of Fleiss's original concepts about human bisexuality were a little awry, his intellectual foresight had pioneered an area of rhythm research that he felt described man's cyclical alterations in both 'masculine' and 'feminine' characteristics.

In 1887, Fleiss published his ideas and findings about human biological rhythms; he considered the twenty-three day rhythm to be fundamentally masculine in nature, and accordingly, he attributed the twenty-eight day rhythm with more feminine qualities. The masculine attributes that he imparted to the shorter cycle were those such as determination, stamina, resilience, and courage. The longer rhythm influenced qualities such as intuition, love, sensitivity, and creativity. Because of Fleiss's theories on the twenty-eight day cycle, it is often called the sensitivity biorhythm. Fleiss proposed that both cycles are apparent in each cell of our bodies, and that they determine our physical and emotional responses throughout our lives. He even proposed that they determine the day that an individual will die. *(See Chapter 6, Rhythm of the Moment.)*

In 1906, Fleiss published *The Course of Life*. This work contained his discoveries about these two cycles that were first investigated through his patients, and together with these revelations he included complex mathematical procedures and tables for the implementation of biorhythm theory. Unfortunately, not only did the general public find his system of calculation far too tedious and daunting, but additionally, his peers in the medical profession considered it too specialized to be of much practical use. Sadly, his literary contribution to the world went, in the main, unnoticed – he was unable to communicate his ideas clearly and effectively to the German nation. Although he was a dedicated expert, he failed to make much of an impression because of the complexity of his delivery.

HERMANN SWOBODA

Dr Hermann Swoboda (1873-1963), a professor of psychology at the University of Vienna became interested in periodic alterations of the mind and body through reading published material that appeared towards the end of the nineteenth century. This stimulated him into starting a systematic program of observation of any physical and emotional conditions in man that were perplexing in their occurrence and timing. He soon began to realize through frequent daily contact with his patients that there was indeed a rhythm in their mental powers of observation and perception, nature of dreams, and emotional states. He also studied the more physical manifestations such as heart attacks, fevers, tissue inflammations, swellings through insect bites, and asthma attacks. In short, he made meticulous and detailed notes about the onset and duration of these various conditions through his eager desire to unfold the mysteries surrounding previously inexplicable, periodic illnesses. Gradually, as he began to accummulate more and more information about his patients, a pattern started to emerge. This pattern showed that there were twenty-three and twenty-eight day cycles in existence for most of the physical and emotional manifestations that he had been able to investigate systematically.

In 1900 he presented a paper to the University of Vienna. Swoboda was fascinated and captivated about the constant 'ebb and flow' that all of us experience – not only in the changing world around us, but also within our own bodies. He considered that although each day can be controlled and structured so that our lives may be conducted in a regular fashion, there still remained the mystery of inner changes that do not seem to have any source in our surrounding environment. His observations started to unfold the very nature of our being: the nature of biological cyclical change. Swoboda had discovered definite rhythms governed by our bodies having an innate sense of time through an internal 'clock' – rhythms that were capable of influencing our response to infections, diseases, stressful conditions, and the daily demands of life. He reasoned that if man was placed in an environment without change, his inner state would alter from day-to-day. Despite all external factors being constant, each of us would experience 'good' days and also 'bad' days.

In 1904, Swoboda published his conclusions about these two inner rhythms. His observations led him to the belief that the twenty-three day rhythm influenced male stamina and sexual potency, and the twenty-eight day rhythm influenced emotional states and degrees of sensitivity in women. It is interesting to note that at this stage in biorhythm awareness, Swoboda and Fleiss were researching independently of each other. They came to amazingly similar conclusions and yet they were totally unaquainted with each other's labours.

As the theory and application of biorhythm began to persuade the medical profession and also the general public, Swoboda realized that the lengthy calculations carried a high risk of computational error for those who were not mathematically minded or used to such lengthy figure work. In view of this, Swoboda designed a slide rule and also wrote a booklet called *The Critical Days of Man* which were made available in 1909. With the aid of this slide rule the whole process of calculating critical days became much easier, resulting in a popularization and enhancement of biorhythm application in his native land of Austria. He also wrote *Periodicity in Man's Life,* which was his first book, followed by *Studies in the Basis of Psychology.* One of his latter works, *The Year of Seven,* his labour of love, came to almost 600 pages, and contained details of his researches into the timing of births: he had discovered a predictable pattern of birth occurrence within families which continued from generation to generation. (An extraordinary present day example of this, which broke all English records, was of a boy and girl born just fifty-three minutes apart on 12 August, 1984. That might not sound very unusual . . . until you realize that Thomas and Jennifer were born from separate mothers and yet, from a medical standpoint, they are brother and sister because their identical twin mothers Pat and Pauline married identical twins John and Peter. The two couples married on the same day, and during the pregnancies both wives developed back pains on the same nights and also had similar cravings for scallops and lobsters.)

Swoboda was indeed a prolific writer, and in addition to his considerable literary output, he lectured at the University informing fellow academics of his advances in the comprehension and application of biorhythm. He also utilized the 'air waves' of Radio Vienna to communicate his findings that this emergent science was a powerful tool for unleashing hidden human potentials. As an adjunct to these studies he also performed analyses on cancerous tissues, and he became noted for his theories on genetic conditions that might propagate malignant cell growth. Even after the Russian invasion of Vienna in 1945 (during which much of his work and statistical evidence were lost), he returned with renewed vigour to his life's work of probing and analysing the human condition. He continued to contribute much to the understanding of biorhythm, and in 1951 the University of Vienna presented him with an honorary degree for the excellence of his advances on physiological rhythms.

Even at this age of seventy-eight (when most of us would feel entitled to a slower pace), Swoboda still continued his research work, and of course his writings that were to form his final contribution to biorhythm: *The Meaning of the Seven Year Rhythm in Human Heredity*. This last book contained further progressions in his observation of birth timing and periodicity for those of common lineage.

ALFRED TELTSCHER

In 1928, Dr Wilhelm Fleiss died, but his life's work on biorhythm continued through the observations and writings of Dr Alfred Teltscher, a doctor of engineering at the University of Innsbruck, and a student of mathematics. By this time, biorhythm science was creating a new awareness of the human mind and body, and this prompted Teltscher to consider investigating this fascinating area of human performance and behaviour.

In the same year, Teltscher commenced his first exploration into our cyclic nature: he obtained details of birth dates, examination dates, and examination performance for about 5,000 Innsbruck high school and college students. His reasoning for this approach to biorhythm research was clear: emotional outlook can be difficult to define at the best of times, and finding statistically valid numbers of subjects for physical tasks can also create problems. Within his own locality he had an ideal opportunity to unravel any mysteries surrounding biorhythm, and a precise model to investigate its claims – the examination of students' intellectual performance by academic experts. All the critical assessment would be done for him, what remained was the statistical analysis of exam results against biorhythm profiles. Unlike Fleiss, he did not require the help of a mathematician as he was already well versed in analysing substantial amounts of numerical data. Because of the large sample of people, Teltscher was confident that if any rhythms existed, they would indeed be apparent. On initial analysis, the data did not reveal any twenty-three and twenty-eight day rhythms. This was hardly surprising, for these two cycles affect behavioural patterns that cannot be detected by the inspection of intellectual performance alone. However, Teltscher did discover that in addition to the periodic patterns described by Swoboda and Fleiss, there was a third, still longer cycle that had the power to influence mental acuity. This rhythm, he discerned, modified the brain's ability to absorb new facts and information, to express ideas in a fluent and effective manner, and also to recall material that had already been assimilated.

Discussions amongst Teltscher's medical colleagues about his discovery of a precise periodicity in students' intellectual performance gave rise to questions about the origin of this mental cycle. Finding a root cause for these biorhythmic phenomena was not a new line of thought, as earlier work on the physical and emotional rhythms had produced definite ideas on the origin of these patterns. For the physical cycle it was thought that our muscle cells give rise to this fluctuation as our musculature controls, to a large extent, our physical abilities and strength. For the emotional rhythm that influences our sensitivity, creative prowess, and mood states, the nervous system was considered a likely source for rhythmicity. (Contemporary research has since revealed that the biochemical balance of the central nervous sytem has a profound effect on our mood states, so initial speculations into physiological sites for the sensitivity rhythm do seem quite acceptable.) The longest rhythm, which Teltscher had shown to influence students' intellectual performance, might well be born within glands that directly affect the brain's deepest workings. Again, modern knowledge of the pineal, pituitary, and thyroid glands indicates that they have a powerful influence on cerebral function.

For the second time, the emergence of biorhythm contained two researchers working independently of, and unaware of each other's endeavours; yet coming to the same interesting conclusions. The discovery of the physical and emotional cycles can be attributed to the near-parallel work of Swoboda and Fleiss; the intellectual cycle can be attributed to Dr Alfred Teltscher and another scientist in Pennsylvania, America.

HERSEY'S RHYTHMIC WORKERS

Over a four year period coinciding with the Innsbruck investigations, Dr Rexford Hersey and his assistant Dr Michael John Bennett examined periodic behaviour in railway and factory workers. During his research at the Pennsylvania University, he collated information from mood questionnaires and interviews with workers' families and fellow employees. The impulse for his work was generated by the general attitude to these employees at the time: many businessmen considered the human 'unit' to be consistent in temperament and output over lengthy periods of time. In fact, the worker was often imputed with more machine-like qualities rather than a realistic assessment of the natural quirks and irregularities that make up our human nature. From the slant of an industrial psychologist, Hersey decided to pursue this concept and evaluate its credibility. Were human beings, he thought, capable of such regular, mechanistic responses day-in and day-out, or did they portray a hidden, undiscovered rhythmicity?

From his many experiments between 1928 and 1932, Hersey found evidence that validated the existence of the Fleiss and Swoboda rhythms. Again, like Teltscher, he discovered a third rhythm that had a periodicity of about thirty-three days. The main thrust of his studies was aimed at appraisal of the emotional nature of man, and surprisingly these probings into human behavioural alterations suggested the longest phasing of all. However, when considering his experimental design, much of the acquired information was obtained by the worker's evaluation and subsequent communication of their inner state of being. Therefore, the mental energy and critical ability of the individual exerted an influence on the degree and quality of expression; thus the intellectual cycle was being inadvertently observed rather than the emotional. (If the experimental design was simply *observing* emotional changes in human subjects, then clearly this regular alteration in mental energy would not be apparent.) Fortuitously, a second, strange coincidence in the discovery of biorhythms had occurred: Hersey's experiments had 'keyed in' to the intellectual cycle simultaneously realized by Teltscher.

JUDT'S ATHLETIC TABLES

From the relatively short interest in biorhythm at the University of Pennsylvania, attention was again focused in Europe, but this time to the North German port of Bremen. As with Teltscher, the destiny of this newly developed discipline depended upon an engineer and mathematician for its advancement and dissemination – Dr Alfred Judt. He decided to simplify the procedures involved in determining a person's biorhythm status by designing tables that would reduce the amount of calculation time. His interest was mainly in the world of sporting activities, and his charts were very much designed for that express purpose. One particular area that attracted his attentions was the application of biorhythm to team performances. But, his efforts had far-reaching effects: apart from being used for looking into athletic accomplishment, Judt's tables were also employed by the medical fraternity to monitor patients illnesses. Although Judt's contribution to this science helped to bring its application into the competitive arena of sport and the exacting domain of medicine, his tables were still too design specific to be of much use to the general public. Biorhythm awareness had taken yet another step in the right direction, but at this stage, a breakthrough in its widespread application to humankind was an aspiration of these early pioneers that was yet to be fulfilled. Further mathematical simplification was still required, and if that goal was achieved, the importance of its benefits had to be understood by ordinary people – not just the medical specialists and mathematicans interested in human behaviour. It was into this uncertain state of affairs in biorhythm infancy that another Swiss engineer injected his practical and mathematical skills that helped to improve the biorhythm 'image' by means of clarification.

FRUEH'S OLYMPIC DESIGNS

Hans Frueh had access to the Judt charts which were designed with the athlete in mind. Upon reflection, he decided that through a thorough revision of these tables, the importance of human periodicity in the physical, emotional, and intellectual cycles could be enhanced. His designing skills went further than this as he created the Bio-Card and also a mechanical device called the Bio-Calculator (which went into production in 1939). The Bio-Card was simply a vertical calculating scale with which individuals were able to review their bioprofiles. For the first time, the general public had access to a system that was simpler and far more elegant than the imposing formulae of the much earlier Fleiss system. Contrastingly, the more up-market Bio-Calculator had the ability to compute a person's wave positions for any required day, and was operated by means of a hand-powered drive. On one side, the calculation tables were displayed, and on the other, alignments for the month of interest. Instead of facing pages of tedious figures, a student of biorhythm could now choose between the Bio-Card scale with the three rhythms – P, E, and I – displayed respectively in blue, red, and green, or the more sophisticated Bio-Calculator. For the Bio-Card, the discharging curves within each cycle were illustrated in their allocated colours, and the negative recharging portions were left blank. This half-representation of each cycle ensured that the critical day crossing points were easily observable; changes of phase were therefore highlighted. With the aid of the Bio-Card many accident investigations were initiated, and once more the sporting world received an affluence of interest. This interest was so embracing that German athletes were trained under biorhythm guidance in preparation for the Berlin Olympics in 1936. Through the efforts of multi-talented Hans Frueh, biorhythm was also being used in the operating theatres of Germany and Switzerland. Biorhythm had reached a point where many surgeons were of one mind in the benefits of timing major operations with cooperative phasing in a patient's rhythms. One surgeon in particular,

Dr Werner Zabel, always advised his patients against surgery (where possible) when biorhythm indicated that an alternative date would be more favourable. After many years of experience with these cycles, he considered that its wide-spread employment would bring innumerable benefits to the world of science. Thanks to Frueh, his Bio-Card and Bio-Calculator, this biological study of rhythms was now leaving its infancy and entering a period of recognition.

NATIONAL AND INTERNATIONAL ATHLETICS

Further advances in athletics were made through the efforts of the Swiss National Coach for Gymnastics: Jack Gunthard. He successfully applied biorhythm forecasting in training and fitness programs for aspiring athletes under his care. His conviction that biorhythm had real and concrete applications in the highly competitive sporting world was evident through the employment of this science in the Gymnastic World Championships in Llublana, Yugoslavia, and also in his native land of Switzerland for the National Championships. He discovered that biorhythm predictions were on average *eighty per cent* accurate for the two climactic finals. In addition to this, he found that some of his athletes responded more markedly to biorhythm planning than others. Through these modified, individual responses, he was able to classify some of his gymnasts as predominantly rhythmic and others as non-rhythmic. With this added tool of biological planning he produced far more accurate assessments about an individual's potential, prior to each day's competition. Thus, his choice of team line-up was more effective. (If he encountered a situation where two gymnasts were of equal fitness and readiness for the approaching contests, he used biorhythm as the 'final word' in arriving at an ultimate choice.)

SWISS ACCIDENTS – THE BREAKTHROUGH

One of Frueh's many disciples was a man called Hans Schwing who, in particular, accomplished a turning point in the study of biorhythm conditions with accidents. His researches were based at the Swiss Federal Institute of Technology under the guidance of Dr W von Gozenbach. This thesis (for a doctorate in natural science) contained details of accident dates and deaths for 1,000 people. Within this dissertation there were two separate studies: the first analysed data from 700 accidents obtained from Swiss insurance companies and the Swiss government; the second investigated 300 mortalities at Zurich, his place of learning. What really interested Schwing was the correlation of biorhythmically critical days with these instances of trauma and death. His results and conclusions were obtained by analysing the ratio of critical/non-critical days that an individual will experience in his lifetime. Hence, it might be helpful to understand some of the initial figure work that formed the basis of his report.

The largest biorhythm unit of time is the *biorhythmic span*. This period is fifty-eight years and sixty-six or sixty-seven days (depending on leap year placements). It is only after this term of 21,252 days that an individual portrays biorhythm conditions that are evident at, and continue from, the moment of birth. Within this span there are a set number of critical days that must occur, and by counting up the total number of these unstable periods and comparing them with the sum of days in the span, a percentage, or ratio, can be obtained. Initially, Schwing approached this problem by calculating the number of single and double critical days that occur within one span for all three biorhythms. He then proceeded further with his analysis and determined the number of single and double critical days in the same period – but only for the physical and emotional rhythms. His reason for omitting the intellectual rhythm in the second calculation was that, at this stage in his research, he was unsure about the precise effect of the intellectually critical day on accident susceptibility and death. By evaluating the two schemes he had constructed a thorough foundation for the interpretation of forthcoming results. These results yielded two values of critical/non-critical ratios expressed as a percentage. Considering all three rhythms, Schwing realized a total of 4,327 unstable days. This worked out to just over twenty per cent. At a few per cent less (for the physical and emotional cycles) his calculations determined about fifteen per cent of our days as critical. Schwing reasoned that if biorhythmically critical days do not present any reason for caution, a sufficiently large sample of accidents will fail to indicate any correlation with this condition. In other words, he concluded that as our lives contain between fifteen and twenty per cent critical days (roughly one in five) a similar percentage of accidents will occur on these days *if* biorhythm theory is incorrect. Conversely, if critical days create a momentary instability for the individual, there should be a significantly higher percentage of accidents and deaths on these days than the predicted amount.

An important point to note is the type of accident that Schwing selected for his research. He only included situations in which the individual was solely responsible for the resultant situation. This meant that the person's biorhythm position was able to be interpreted without the additional confusion of having other external factors impinge upon the predicament, e.g. mechanical failure, inclement weather, or the actions of a third party. He excluded all instances that demonstrated any hint of 'interference'.

Next came the most demanding part of all: the analysis of each accident in terms of the person's birth date, incident date, and biorhythm position. For Schwing (as for Teltscher), there was no short cut that could be made in the acquirement of biorhythm statistics. A total of one thousand case histories had to be evaluated, classified, and finally interpreted. This considerable effort seems a daunting task, especially in the light of today's computer assisted research. But, Schwing was not alone in his labours, for his style of research set the trend for future scientific work. Retrospective analysis of large numbers of accident case histories is tedious and repetitive, but as he soon discovered, the revelations and implications of biorhythm were quite astounding.

Taking first the 700 accidents, Schwing calculated that 401 had materialized on critical days. A quick, finger-tapping burst on your pocket calculator will reveal that this is way off the fifteen to twenty per cent range: it works out to about fifty-seven per cent – $401 \times 100/700 = 57.28\%$. This percentage is remarkable in itself, but Schwing went still further with these figures and determined that this set of data illustrates that a person is *five times* more likely to have a traumatic incident on critical days as on non-critical (mixed rhythm) days! His findings for the three hundred mortalities were even more significant. Of these deaths, 197 occurred on critical days. Similarly, a quick calculation reveals that about sixty-five per cent of deaths happened when the person was experiencing an alteration of phase from minus to plus or vice versa. By further calculation, these results show us that a person is *eleven times* more likely to die when passing through a critical day! It is no surprise that this dissertation was instrumental in encouraging the development of additional research projects. His eighty-seven page thesis was earmarked as the keystone for further European investigation after the end of World War II.

JAPANESE CHILDREN AT RISK

Five years after Kitchinosuke Tatai received George Thommen's book on biorhythm, its repercussions had spread far enough within the capital for the Metropolitan Police to investigate traffic accidents in high-risk Tokyo. After accumulating statistics for about a year, they published a report in 1971 giving details of various highway incidents. They found that *eighty* per cent of accidents occurred on the driver's critical day! This promptly generated further investigations, and one in particular researched into the susceptibility of children to automobile accidents whilst passing through unstable critical periods. Again, the correlation was alarmingly high. It was discovered that *eighty-five* to *ninety* per cent of accidents involving pedestrian children under the age of ten occurrred when the victim was experiencing a critical day. Other traffic accident data were analysed using the rapid Biocom 200 biorhythm-card computer; the results were consistently high. One report that examined traffic accidents within the districts of Chiba and Kanegawa yielded the same findings as the Tokyo Metropolitan Police – *eighty* per cent. Although the high percentages of driver criticality with incident date makes sense when viewed against biorhythm theory, the results from analysing trauma involving pedestrian children under ten years old does seem, on first sight, to be puzzling.

Shirai – who first instigated this research plan into younger pedestrians – realized that individual behavioural patterns undergo marked changes with increasing age due to societal and parental conditioning. These changes (when considering road safety) are brought about through an increased awareness of potentially dangerous situations. For instance, a five-year-old will have a very different impression of highway hazards to a ten-year-old. He reasoned that this early 'innocence' of children leaves them far more responsive to their critical phases as there is precious little, or no, inhibiting factor that is evident in the more mature individual. This high correlation of children's critical days with car accidents does indicate that we must spend a lot more time explaining the dangers of highway traffic to our young ones. What is true for Japanese children is true for children of other countries. Only through a systematic and thorough instruction of road safety *and* biorhythm awareness will we be able to instil into our youngsters a sense of proper caution when they intend to cross the road.

AT BASE, IN THE FIELD, AND IN THE ROAD

In their enthusiasm towards biorhythm, the Japanese left very few stones unturned in their zealous efforts to reduce traffic accident occurrences. One of these previously unexplored territories was the military establishments, and it was the Kasuga base of the Military Police that came under the scrutinizing, biorhythmic 'eye'. Over one thousand accidents were sifted through for instances of driver criticality. Biorhythm revealed that not even these highly trained, extremely fit specimens were exempt from the cycles of life. The Kasuga patrolmen exhibited critical days in *fifty-nine* per cent of all accidents. (Remember that critical days only account for about twenty per cent of our days, so even this level of incidence is remarkable.) Other classifications of accident that involved higher degrees of driver judgement error yielded results up to sixty-six per cent. Similar results in this sixty per cent range can be found in studies of drivers from other countries, but this time from those that work on the land rather than the defenders of it.

Returning again to Berlin, where biorhythm first came into being through the researches of Fleiss, the mid-1950s witnessed further work by a researcher at the Humbolt University: Dr Reinhold Bochow. His studies concentrated on drivers engaged in working with agricultural equipment and the report encompasses almost five hundred case histories of accident. Although the individuals were of a wide age range (and so they had varying degrees of experience), they were well trained and acquainted with machine operation and the intricacies of the job in hand. All the accidents that occurred were 'accidents': situations where an individual portrays a momentary lapse of concentration rather than downright carelessness. Bochow's results revealed that *sixty-two* per cent of the incidents occurred when the driver was critical.

Another investigation into driving accidents was carried out on the other side of the globe, in Australia. This time, the usual analysis of incidents on the critical day was performed, but in addition, Krause-Poray also examined the day before and the day after a critical event for further information on the nature of our responses adjoining these unstable periods. From his sample of one hundred accidents he discovered that *fifty-four* per cent of fatalities involving the driver, aligned with their critical day. When the days either side of the critical day were also taken into account, this result was increased to *seventy-nine* per cent. This extra twenty-five percent illustrated a common phenomena in statistical analysis known as distribution.

Accident statistics reveal that although the majority of traumas happen on critical days, there is an expected spread of results that often characterizes data from biological systems. Krause-Poray's findings reinforce this principle in that there is a higher than chance occurrence of driver fatality on these adjoining days (called semi-caution days), although far less than the expected peak on the central critical day.

JAPAN IN THE 1960s

Coming forward by a decade from the agricultural experiments of Bochow, the Japanese again predominate with their vigilant analysis of traffic accidents, but this time on a larger scale by a transportation firm that operated fleets of buses and taxis – the Ohmi Railway Company. Because of Japan's high population density, traffic problems are particularly severe. This had a definite bearing on the development and application of biorhythm in the 1960s. Many of these reports were then seriously considered by Western countries for their contribution to possible causes of road accidents and fatalities.

The Ohmi Railway considered accidents over a period from 1963 to 1968. Like the Australian experiments of Krause-Poray, the days either side of the critical day were also taken into account as high risk days. Over these five years, Ohmi discovered that *fifty-nine* per cent of the accidents occurred on the driver's critical day. Analysing the data still further, they discovered that the critical days of more accident prone drivers accounted for an even higher percentage of incident by an additional two per cent. This raised level does bring home the point that some personalities are more influenced by phasing than others. The accident prone individual (perhaps for reasons of deep-seated nervousness) is more susceptible to the momentary changes of rhythm from one phase to another. When the three cycles were considered in turn, Ohmi found that the physical cycle precipitated a greater number of incidents than the other two rhythms. And, when double critical days had been analysed, the physical/emotional had the most disturbing effect of all.

Driving really needs special attention in itself, for it is an extremely complex procedure that most of us take for granted. For many, this daily habituation increases our road experience and subsequent competence, but familiarity can also germinate the seeds of carelessness that are so often evident at the scene of misfortune. Slowed reactions when driving too close, lapses of concentration when pulling out to overtake, ruffled nerves due to an argument with a back seat driver, all contribute more than amply to conditions that culture hazardous road situations. In type these represent the physical, intellectual, and emotional realms of behaviour that can be modified by our biorhythm position.

After the lengthy five year period of collating accident data that vindicated the relevance of biorhythm, Ohmi took this neo-Fleissian system one step further in its beneficial convergence on the human condition. This step was powerful, simple, and effective. When a driver was approaching an unstable critical event, they issued him with a warning card that hailed his biorhythm status. The results were dramatic: in the first year of development, their personal assault on needless human suffering had triumphantly slashed accident rates by *fifty* per cent. Successive years announced further reduction, and after four years, they had managed to reach an unprecedented level of

28

competence: *four million* kilometers on the roads without a single collision. This record is truly a feather in the biorhythm cap, but when one appreciates that traffic density was on the increase, together with worsening accident statistics in Japan as a whole, it becomes all the more astonishing. This is by no means a single, isolated example of biorhythm application and planning in road safety schemes, for many other commercial and industrial businesses in Japan have found out that ignoring biorhythm means ignoring road death.

Similar studies have been carried out in the cities of Hanover, Germany, and Zurich, Switzerland, where bus and tram drivers were warned of their biorhythm instability when phase changes were about to take place. With the obvious high mileage that this type of business encounters each year, accident statistics become that more meaningful. Records of the Zurich Municipal Transit Company have shown that within one year of commencing this scheme, their accident rate plummeted by *fifty* per cent – identical to those reported by the Ohmi Railroad Company.

This period in biorhythm history saw intense activity in road safety schemes, but there was also a resurgence of medical interest in the treatment of illness and disease that had initiated those pioneering explorations into body rhythms in Berlin and Vienna.

2 THE CALCULATIONS

'Is this what science teaches us?' asked Mr Tompkins with interest. 'On these great problems science is not sure what it teaches us,' replied the stone deity. 'Our universe, your scientists think, had a beginning in time. Something blew up; it formed the universe and kept expanding. Perhaps it will keep going and eventually vanish into nothingness, or perhaps it will collapse again into a primeval atom, blow up again, and keep repeating itself indefinitely. They do not know, at least not yet. But at any rate, without knowing the answer to these great problems, we measure time by the recurrence of events.'

Mr Tompkins Inside Himself – George Gamow and Martynas Ycas

In this chapter we will learn to measure the time of our own bodies; dividing our days by the three rhythms that have an ability to influence our physical, emotional, and intellectual powers. Through a daily knowledge of your biorhythms a new awareness will manifest itself. An awareness that is essential, for we are physical systems – biological 'clocks' in a precisely timed world. Through understanding the process of calculating anyone's biorhythm position, you will be able to realize new potentials in yourself, and also in those with whom you have frequent daily contact. Biorhythms are not just for individuals, but can be used to improve relationships within groups as well. Apply your knowledge of biorhythms to your work situation. If relationships are a bit strained at times, take the trouble to find out your biorhythm position, and also the condition of those around you. Perhaps you live in a large family; take your understanding of these three inner cycles right into your home. Be creative and improve the quality of your life when socializing with friends. There is no limit to what you will be able to achieve. Where there are people, there are living 'clocks'. Each of us has good days and bad days. Only through knowing your next critical day, high phase, or low phase, will you be able to plan constructively in a constantly changing world. Above all, take the time to apply what you know. So, now to explain the method for determining your rhythms with a calculator and how to draw your own biocharts.

USING A CALCULATOR

You may feel a bit overwhelmed by the prospect of having to play around with masses of figures but this section will reassure you that although the procedure is a bit complicated, it can be easily understood and applied, provided it is followed carefully and methodically. Practice makes perfect. I have given three detailed examples of how to work out someone's biorhythmic make-up by using the indicated method. Follow the examples carefully and soon you will be doing it by rote. At the back of the book you will find tables that will help you in the calculation process.

In order to obtain the positions for each of your biorhythms on any given day, you will have to work out all the days that you have lived, from the day you were born up to the date that you are interested in analysing (inclusively). This will be the first day of the month for biocharts. The detailed figure work following these three examples will help you to 'get to grips' with the mathematical process:

PERSON 1 Jennifer Brown, who was born on 8 February 1966, wants to know her biorhythm position for July 1986 as she has decided to take a two week holiday abroad at some point in that month. She wonders if her biorhythms will be able to indicate the best dates for her vacation. There will be the inevitable change of diet and climate, together with a shift in time zones of nine hours. Her hobbies include horse riding and walking, and she prefers out-of-doors holidays that cater for the more active type of person. All the usual changes when travelling and spending time in another country will have to be adjusted to. Will her physical rhythm be in a suitable phase to cope with all these simultaneous alterations? There are also the emotional and intellectual cycles to consider as well. Will she be at an emotionally low phase during this two week holiday? If so, then this biorhythm condition must be remembered and compensated for. Will she have an intellectually critical day just prior to travelling? If this does occur then preparations must be made in order that nothing will be forgotten; planning will be a very wise precaution here.

PERSON 2 Michael Hanson, who was born on 14 October 1964, wants to know how he will fare in a long distance swimming competition on 12 April 1986. If his physical rhythm is high, this will hold him in good stead, although an emotional high might well lead to being over confident with subsequent changes in his behaviour during the swim. This condition is a possibility because, like Jennifer, Michael is an extrovert and is easily excited: his biorhythm peaks need careful planning and evaluation. An intellectually critical day would also require attention as decisions made during the tournament could be affected by this temporary instability in mental skills. General biorhythmic trends during the week before the swim will be of assistance in guiding his work-outs and rest schedules by following the physical and emotional cycles.

PERSON 3 Simon Watkins, who was born on 8 August 1957, has developed a keen interest in sky diving. Over the last year he has put in a lot of practice and engaged in numerous jumps. For his twenty-ninth birthday, his friends would like to arrange a special jump with him, and at the same time try out a new formation in the air. Considering the excitement of having a birthday, and celebrating it in this most unusual way, can biorhythm analysis give any indication if this imaginitive birthday gesture will be a success? All three rhythms come into play when considering this exhilarating sport. The physical rhythm controls muscle coordination which is vital for mid-air link-ups. Stamina isn't so important as the dive duration is relatively short, but the ability to grip well is a must. The emotional rhythm might play a very important part on 8 August, for if Simon is easily excitable and his emotional rhythm is high, then an accident might well result from his level-headed approach being momentarily out of balance. The intellectual cycle involves decision-making ability, and in negative and critical phases it reduces our concentration, which becomes more apparent in mentally demanding situations. During the jump, Simon's brain will be constantly analysing his changing environment as his collegues proceed with the dive. Critical days should therefore be avoided.

Now for the figure work which will give us some answers for Jennifer, Michael, and Simon. If the analysis date and the birth date are less than one year apart, start the procedure from step four. The stages in calculating the number of days from the birth date to the analysis date are as follows:

1   If the analysis month falls later in the year than the birth month, subtract the analysis year from the birth year, e.g.

JENNIFER   1986 − 1966 = 20 years
MICHAEL   does not apply
SIMON   does not apply

2   If the analysis month is the same as, or falls earlier in the year than, the birth month, subtract the analysis year from the birth year, and subtract one again, e.g.

JENNIFER   does not apply
MICHAEL   1986 − 1964 − 1 = 21 years
SIMON   1986 − 1957 − 1 = 28 years

33

3   Multiply the result obtained in (1) or (2) by 365 to obtain the number of days during this period, e.g.

JENNIFER $20 \times 365 = 7,300$ days
MICHAEL $21 \times 365 = 7,665$ days
SIMON $28 \times 365 = 10,220$ days

4   The calculation in steps (1) or (2) will give you the number of days from the birth date to the preceding calendar month date, e.g. In Jennifer's case, 8 February, 1966, to 7 February, 1986. In this step, count the days from this date to the end of the last month by subtracting the date from the number of days in that month. Refer to the *Table of Days in each Month* at the back of the book, e.g.

JENNIFER $28 - 7 = 21$ days
MICHAEL $31 - 13 = 18$ days
SIMON $31 - 7 = 24$ days

5   Add up all the days for all subsequent months up to the month *before* the analysis month, e.g. in Michael's case, November through to March. Refer to the *Table of Days in each Month* at the back of the book, e.g.

JENNIFER $31 + 30 + 31 + 30 = 122$ days
MICHAEL $30 + 31 + 31 + 28 + 31 = 151$ days
SIMON $30 + 31 + 30 + 31 + 31 + 28 + 31 + 30 + 31 + 30 + 31 = 334$ days

6   Take into account any leap days that have occurred during the period from the date of birth to the date of analysis. Refer to the *Table of Leap Years* at the back of the book, e.g.

JENNIFER   1966 to 1986 = 5 leap days
MICHAEL   1964 to 1986 = 5 leap days
SIMON   1957 to 1986 = 7 leap days

7   Add together the number of days obtained in (3) – if applicable, (4), (5), and (6). Then add one to give the biorhythm position for the first of the month (the analysis month), e.g.

JENNIFER $7,300 + 21 + 122 + 5 + 1 = 7,449$ days
MICHAEL $7,665 + 18 + 151 + 5 + 1 = 7,840$ days
SIMON $10,220 + 24 + 334 + 7 + 1 = 10,586$ days

8   The number obtained in (7) is the total number of days from the date of birth to the date of analysis (usually the first of the month for biocharts). This number must then be divided by the lengths of each of the three biorhythm cycles, i.e. physical cycle = 23 days, emotional cycle = 28 days, intellectual cycle = 33 days, e.g.

JENNIFER   Physical $7,449/23 = 323.86956$
                Emotional $7,449/28 = 266.03571$
                Intellectual $7,449/33 = 225.72727$

MICHAEL	Physical	$7,840/23 = 340.86956$
	Emotional	$7,840/28 = 280$
	Intellectual	$7,840/33 = 237.57575$
SIMON	Physical	$10,586/23 = 460.26086$
	Emotional	$10,586/28 = 378.07142$
	Intellectual	$10,586/33 = 320.78787$

Usually with this calculation you will get several numbers after the decimal point, but these can be ignored. *Reduce* the number to the nearest whole number if necessary.

9 Multiply the whole number obtained in (8) by the respective number of days it takes each cycle to repeat itself, e.g.

JENNIFER	Physical	$323 \times 23 = 7,429$ days
	Emotional	$266 \times 28 = 7,448$ days
	Intellectual	$225 \times 33 = 7,425$ days
MICHAEL	Physical	$340 \times 23 = 7,820$ days
	Emotional	$280 \times 28 = 7,840$ days
	Intellectual	$237 \times 33 = 7,821$ days
SIMON	Physical	$460 \times 23 = 10,580$ days
	Emotional	$378 \times 28 = 10,584$ days
	Intellectual	$320 \times 33 = 10,560$ days

10 This is the last and final stage, you're almost there! Subtract from the total number of days – the figure obtained in (7) – the three numbers obtained in (9). These final numbers will give you the phase position for each of the three cycles on the day of analysis, e.g.

JENNIFER	Physical biostart	$7,449 - 7,429 = 20$
	Emotional biostart	$7,449 - 7,448 = 1$
	Intellectual biostart	$7,449 - 7,425 = 24$
MICHAEL	Physical biostart	$7,840 - 7,820 = 20$
	Emotional biostart	$7,840 - 7,840 = 0$
	Intellectual biostart	$7,840 - 7,821 = 19$
SIMON	Physical biostart	$10,586 - 10,580 = 6$
	Emotional biostart	$10,586 - 10,584 = 2$
	Intellectual biostart	$10,586 - 10,560 = 26$

At the back of the book you will find a table headed *Blank Biochart Day Count*, which you can photocopy and fill out to help you through the process. The ten steps that are needed to complete the day count are similarly numbered on the table. In addition to this I have included a reminder of the procedure within each step, thus avoiding frequent referral to this chapter. The following illustration shows the table filled out for Jennifer.

35

NAME	J. BROWN		DOB	8.2.66
			MONTH	JULY '86
1	YEARS		=	20
2	YEARS		=	
3	(1) OR (2) × 365		=	7300
4	DAYS TO END MONTH		=	21
5	DAYS IN MONTHS		=	122
6	LEAP DAYS		=	5
7	TOTAL OF (3),(4),(5),(6) + 1		=	7449
8	(7) / 23	(P)	=	323.86956
	(7) / 28	(E)	=	266.03571
	(7) / 33	(I)	=	225.72727
9	(8P) × 23	(P)	=	7429
	(8E) × 28	(E)	=	7448
	(8I) × 33	(I)	=	7425
10	(7) − (9P)	(P)	=	20
	(7) − (9E)	(E)	=	1
	(7) − (9I)	(I)	=	24

If in the process of calculating the biostart numbers a zero is obtained (as in Michael's emotional rhythm), convert the zero to the highest number of that cycle, i.e.

Physical biostart 0 converts to 23
Emotional biostart 0 converts to 28
Intellectual biostart 0 converts to 33

This situation always occurs if the total number of days counted equals an exact number of cycles for any of the biorhythms. This conversion is the only exception in the process; any other biostart numbers can be left as calculated. All the hard work has now been done, so if you have confidently sailed through all those computations, I am happy to say that biorhythm theory has thrown its worst at you!

Now that you have calculated the three biostart numbers for the person of your choice, please make a careful note of them as they will save you a lot of hard work for consecutive biocharts. Buy a notebook for all your friends' birth dates and computed biostart numbers for the first month that you will be analysing. By commencing your interest in biorhythms methodically, you will save yourself endless hours of frustration and annoyance through mislaid figure work. If you intend to chart a large number of people, cultivating orderly habits will increase your confidence in the process – it is only too easy to make mistakes.

The illustration represents Jennifer's biorhythm condition for July 1986. As mentioned earlier in the chapter, she would like to take a two week holiday sometime during this month. By looking at her chart, have you any ideas about which would be a suitable start date for her adventure abroad? The first step is to examine the chart in more detail, looking for general trends and critical day occurrences in the light of her planned activities for this month.

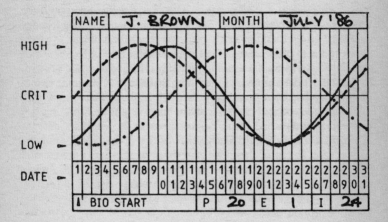

Starting firstly with the physical rhythm, there are critical days on day five, midnight of the sixteenth/seventeenth, and the twenty-eighth. The discharging phase or positive phase spans from the fifth to the sixteenth, followed by a balancing, recharging phase from the seventeenth to the twenty-eighth.

Considering the emotional cycle, there are critical days on the first, fifteenth, and twenty-ninth. The positive and negative conditions of this cycle follow the physical cycle quite closely, so there is an 'additive' effect in operation between these two rhythms for this summer month. One noticeable feature is that the physical and emotional cycles are both negative from the seventeenth to the twenty-seventh. Besides this, there are critical days in each cycle only one day apart towards the end of the month: the twenty-eighth and twenty-ninth.

For the intellectual cycle there are only two critical days (the maximum that can occur in any month), and these appear on the eleventh and midnight of the twenty-seventh/twenty-eighth. The negative phase of this rhythm occupies the first eleven days of the month, and also the last four; this produces rather an unwelcome bunch of critical days at the end of the month – and all within two days. This condition can be described as a double critical day between the intellectual and physical rhythms, followed by a critical day in the emotional rhythm.

All these biorhythmic trends and critical day events have to be viewed against possible physical, emotional, and intellectual 'loads'. When considering a holiday, there will be a greater 'load' on the physical resources of the body whilst travelling (especially in Jennifer's case, with a nine hour time shift), so the outward and return dates should be free of critical days if at all possible. This is especially important if the traveller is also driving to his or her destination: that extra long journey coupled with a critical day vastly increases the chances of having a road accident due to phase changes and the inevitable fatigue factor. If this cannot be entertained then an intellectually critical day would produce the least potential disturbance providing that adequate planning before the journey has been done. Any 'woolly' thinking when faced with unknown connections between trains or buses might well spell disaster before the hard-earned break begins. Looking at Jennifer's chart it becomes clear that outward journeys on the first or fourteenth would be very unwise. Either of these two start dates would mean outward and return journeys on a critical day, or within one day of it (called a semi-caution day). The double critical day on the twenty-eighth followed by a critical day on the twenty-ninth would be a most inauspicious time for a long return journey.

Apart from planning so that a critical day does not coincide with a major journey, lows in the physical and emotional cycles should also be carefully considered. (Each person reacts differently in the recharging and discharging phases – it is here that biorhythmic awareness comes into its own. Through a knowledge of your predominant reactions during each cycle condition, future planning will become much more effective.) However, laying aside specific individual reactions, the general trend is for a lower energy condition to exist in the negative span. The simultaneous physical and emotional lows during the period already indicated does reveal that a holiday starting in the middle of July would be less enjoyed as the body and personality is unable to respond more actively in the physical and emotional areas. This double low would definitely dampen her spirits a bit; so at this stage in the analysis, travelling at the beginning of the month, and spending a vacation over the latter half of the month are not advised.

Because the physical and emotional rhythms are reasonably close to each other during July, there are corresponding advantages in the first half of the month. During the initial fifteen days, critical events are evenly spaced, being on average about five days apart – a much better situation than the near double critical in the middle of the month, and the total phase changes around the twenty-eighth/twenty-ninth. The physical and emotional highs also mean that Jennifer will have plenty of energy to pursue all the activities that she desires, together with the added bonus of enjoying an emotionally positive phase. This will strengthen her extrovert qualities, thus improving her approach to newly formed friends and the unfamiliar surroundings abroad. Avoiding the physically critical day on the fifth for a travelling day, a sensible choice for the outward journey would be on the fourth, or possibly the sixth. The return dates on the eighteenth or twentieth would also miss any critical days, although the best choice is the earlier. Latter dates for outward travel from the seventh to the tenth could also be entertained, but this will then include the necessary depression in the emotional and physical cycles during the latter half of the holiday.

By looking at each cycle individually, and then the biochart as a whole, it is possible to arrive at a suitable compromise for any planned activities. As previously mentioned, a knowledge of your personality responses and physical energy trends during each biorhythmic phase will greatly aid in interpretating your personal biochart.

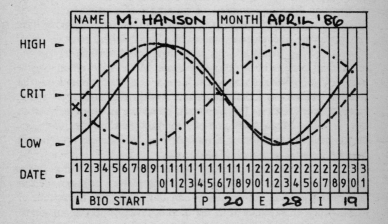

This illustration represents Michael's biorhythm condition for April 1986. The interpretation of this chart is much simpler than Jennifer's due to the interest in one particular date rather than a more general analysis of the whole month. The date for the swimming tournament is 12 April. From the chart it can be seen that on this day the physical and emotional cycles are high, but the intellectual rhythm is low. At first glance this is an ideal combination for a physical activity such as swimming where a lot of stamina is required – especially as this is a long distance swim. However, the emotional rhythm needs more careful consideration in the light of personality types.

Biorhythm research has shown that the emotional rhythm does play a large part in our physical response to goal-seeking tasks. The reason for this is clear: our emotional attitudes will colour our motivation which in turn influences our physical reactions. For an individual who is basically introverted, emotionally low periods may well need special attention due to a low energy condition inhibiting someone who may have difficulty in developing a more outward approach. For an extrovert who operates on a more socially interactive basis, and who would tend to be a high reactor type, emotional highs do require caution for exactly the opposite reason. In a sense, an emotional peak for a high reactor could introduce conditions in which it would be easy to 'burn out' through an over exuberant approach to the task in hand. When some people are really 'high' about something, it is all to easy for them to exert the body to a point beyond fatigue, thus causing stress, and if continued, trauma.

Michael is a high reactor type, and so there will be a tendency for him to over respond during the long distance swim. Because this type of competition involves stamina, and training over a long period, much could be lost due to swimming too strongly at the start and leaving little energy for the final stages of the swim. This is where good coaching plays a vital part in any athletes performance, and considering that the intellectual cycle is in a recharging phase, Michael is well advised to listen to the expertise of his trainer. Here, a 'cool', calm approach to the swim is required as a definite balance against the tendency towards physical over-exertion. The important points are an even distribution of energy during the swim and not going in with too much 'attack', coupled with taking heed of any advice from his trainer on the day, in view of the intellectually low phase.

The illustration on page 41 shows Simon's biorhythm wave conditions for August 1986. Again, in this analysis only one date is under consideration: Simon's birth date, 8 August. The birthday treat is for Simon to sky dive with his friends, and also to try out a new formation in the air. Undoubtedly, this would be an exciting birthday gesture and a wonderful way to celebrate. Does biorhythm theory counsel any changes in the plans of his sky diving collegues? Would a rescheduling of the dive to another day be advised?

Examining the physical rhythm, it appears that a change of phase will occur at about midnight on the seventh/eighth, and that the cycle will

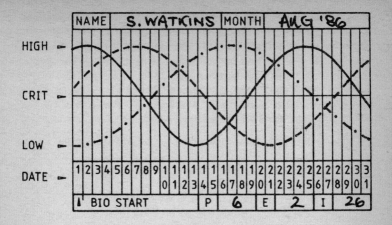

| NAME | S. WATKINS | MONTH | AUG '86 |

DATE	1	2	3	4	5	6	7	8	9	10	11	12	13	14	15	16	17	18	19	20	21	22	23	24	25	26	27	28	29	30	31

1 BIO START   P 6 E 2 I 26

be recharging after that point. The emotional rhythm is high on the
planned day for the jump, in fact this condition can almost be called a
mini-critical day – when any rhythm is at its highest or lowest point.
The intellectual rhythm is critical on the ninth, and because this cycle
repeats itself at a slower frequency than the other two, the critical day
has a slightly broader influence each side of the crossing point. (For the
emotional and physical cycles, one day either side of the critical day is
sufficient; these are called semi-caution days due to a higher occurrence
of incidents on these days. For the intellectual biorhythm, the critical
day centers on a period of semi-caution that lasts for about three days.)

To sum up, the planned jump coincides within one day of physically
and intellectually critical days, but also aligns within one day of the
highest point in the emotional rhythm. In view of these conditions,
perhaps a more detailed look at the effect of an emotional peak coupled
with these two phase changes might reveal probable outcomes in what
is considered a high risk sport, rather than viewing the two criticals in
isolation.

In Simon's case, a physically critical day and an emotional peak are
only one day from the planned birthday event. On days when the
physical rhythm is changing phase from positive to negative, physical
reactions will be much slower than usual. There is also a tendency for
greater fatigue when the discharging phase is terminated. This lack of
stability in muscle coordination, together with increased fatigue must
then be considered with the added factor of an emotional high.
Potentially, this situation could introduce very high risks into the
jump, especially as a new formation is being tried out. That will require
extra concentration and Simon is within the influence of an

intellectually critical day. The emotional high will tend to propagate feelings of well-being and stability, yet 'underneath' the body will be changing phase, and the mind will be following the same reversal of polarity. The conclusion is that Simon is definitely advised against participating in this birthday event, even though he might 'feel' that all is well and that nothing could go wrong. That is where the danger lies in this particular biorhythmic combination – when we *feel* that everything is safe and going smoothly.

All of us have had times when we felt that all was OK, perhaps being too confident in our own bodies or abilities. Biorhythm theory does instruct us to be watchful when passing through emotional peaks. Each of us need to evaluate our own personalities and see if there is any likelihood that we will react in an over-confident manner during these conditions. For Simon, the answer has been a definite no. A better date would be around the twelfth. The intellectual high would enable him to concentrate on the new formation, giving an extra margin of safety. The physical low might initially seem to be a very bad choice, but this is where fitness and past experience comes into play. This is a better choice of date than an earlier one such as the sixth. In this condition, the intellectual low is really counter productive in learning the new diving formation, and the physical high, although provident, is better sacrificied for the wave positions just a few days later.

DRAWING A BIOCHART

The next stage in the process is drawing your biocharts. If you turn to the end of the book, you will find blank biocharts and also three bioguides (one for each of the cycles). The easiest part is the photocopying of the blank biocharts and the three bioguides. Take more copies of the biocharts than you need so that you can practise drawing the wave forms and aligning the guides before attempting the real thing. Take more than one copy from each of the bioguides as you will be sticking them onto cardboard. If you should make any mistakes when cutting along the outlines, all will not be lost on your first attempt.

A brief description of the biocharts and bioguides will help to explain the method. The biocharts have been designed so that all the relevant information is contained on the chart. If you develop the habit of filling in all the blanks, future work will be much simpler.

The top row of the biochart contains spaces for the name of the person charted, and also the month being analysed – remember to write in the year as well. (If several past events and incidents over quite a time span are being researched for one individual, this will save confusion when reviewing your work.) The construction of the main part of the biochart is similar to others that are commercially available. The chart is divided into thirty-one days, with high, low, and critical points in the cycles indicated by arrows at the left-hand-side. The dates have been offset to the bottom of the graph in order to increase clarity when observing critical day crossing points. The vertical lines either side of the dates relate to a time span from midnight of the current day to midnight of the next, i.e. a mid-point between the vertical lines equals midday on that date.

The bioguides contain a calibrated scale along the lower edge. With the aid of your computed biostart numbers each bioguide can be easily positioned on the biochart and then drawn with the colour indicated. You can, of course, use what colours you wish, but the three chosen are the internationally accepted ones. If you are charting just for your own use then this really doesn't matter, but having some standards when producing biocharts for other people will help to communicate biorhythm theory more effectively by minimising confusion.

Now onto the practical aspect of constructing the three bioguides. Initially, I made a few mistakes in the process, so I will pass on a few tips that will smooth your path to successful charting. The materials that you will need are as follows:

A scalpel or sharp 'Stanley' knife.
A cutting board.
A steel rule – one foot.
A suitable glue for sticking paper to cardboard.
Cardboard – about one foot square.

This list need not be very expensive. If you cannot get hold of a scalpel then it should be possible to obtain a sharp-bladed hobbyist's knife that will serve the same purpose. A blade that is pointed will be able to follow the curved cycle line more easily than one with a broader profile. The cutting board can be some hardboard of a reasonable thickness – the type that most artists would use for oil painting when suitably prepared. The steel rule is advisable as a wooden or plastic one would soon lose its edge when guiding a very sharp blade. The knife could also cut through a rule of this inferior type if too much pressure was employed, with the result that the cutting edge might come dangerously close to your fingers. For the cardboard, I experimented initially with quite a heavyweight board with the intention of making the bioguides as durable as possible. This I found to be an unwise choice as it was very difficult, even with a scalpel, to cut through the card in one go. This might not seem to be a problem with straight edges as the steel rule will always hold the blade to the same position. However, when cutting through the curved, dotted line, several cuts tended to reduce the chance of obtaining a clean, perpendicular edge. This I found to be the main source of inaccuracy when cutting through this part of the bioguide. Cardboard from a breakfast cereal packet is about the right thickness as it is possible to go through with one cut. For those of you who are experienced in using sharp-bladed knifes, you might like to skip the following safety list of Do's and Don't's:

Do use a cutting board at all times.
Do work on a solid level surface when cutting.
Do make sure that you have adequate illumination.
Do use the steel rule rather than an inferior type.
Do finish by covering the blade, and store in a safe place away from pets and children.
Don't apply too much pressure on the blade as it may snap.
Don't finish the cutting stroke with the blade too near your body.
Don't hold the knife whilst using your hands for other manoeuvers.

When cutting out the bioguides, cut fractionally to the outside of the lines and dots rather than through them. With the finished template, all lines should be visible. If it appears to be oversize, you can always trim later. When cutting through the dotted line, take as much time as possible: your accuracy in this step will ensure equally accurate biorhythm forecasting.

The bioguides are designed so that the bottom edge is butted against the line above the biostart row at the bottom of the biochart. When properly located, the date row on the biochart will not be visible. Simply look at the appropriate biostart number in the bottom row of the biochart and slide the bioguide to the left or right until the arrow in the open box (bottom left) corresponds with the correct number on the guide. The line to the left of the biostart number on the bioguide must correspond with the left edge of the biochart. The short line to the right of the arrow on the biochart should also align with the line to the right of the biostart number on the bioguide. All that remains is for you to draw in the wave form with the correct colour. Describing the procedure might seem a bit complicated, but when followed through in practice, it's simple.

If you would like to check the accuracy of your bioguides, the following routine will indicate if there are any cutting irregularities:

1 PHYSICAL BIOGUIDE Align the bioguide for a biostart number of one, and draw a line from day one to day thirty-one on the biochart. The first crossing of the critical line should be half way through day one. The second

crossing of the critical line should be the line between day twelve and day thirteen. The third crossing of the critical line should be half way through day twenty-four. Align the bioguide for a biostart number of twenty-three, and draw a line from day one to day thirty-one on the biochart. The first crossing of the critical line should be half way through day two. The second crossing of the critical line should be the line between day thirteen and day fourteen. The third crossing of the critical line should be half way through day twenty-five.

2 EMOTIONAL BIOGUIDE Align the bioguide for a biostart number of one, and draw a line from day one to day thirty-one on the biochart. The first crossing of the critical line should be half way through day one. The second crossing of the critical line should be half way through day fifteen. The third crossing of the critical line should be half way through day twenty-nine. Align the bioguide for a biostart number of twenty-eight, and draw a line from day one to day thirty-one on the biochart. The first crossing of the critical line should be half way through day two. The second crossing of critical line should be half way through day sixteen. The third crossing of the critical line should be half way through day thirty.

3 INTELLECTUAL BIOGUIDE Align the bioguide for a biostart number of one, and draw a line from day one to day thirty-one on the biochart. The first crossing of the critical line should be half way through day one. The second crossing of the critical line should be the line between day seventeen and day eighteen. Align the bioguide for a biostart number of thirty-three, and draw a line from day one to day thirty-one on the biochart. The first crossing of the critical line should be half way through day two. The second crossing of the critical line should be the line between day eighteen and day nineteen.

If all the critical day crossing points agree with those indicated, the bioguides are ready for use in constructing your biocharts. When drawing along the three bioguides, remember to draw only as far as the last day in the month. With this added visual cue, it will be easier to remember to add the right number of days to the three biostart numbers for their correct values in the next month. The simple procedure that follows is all that you have to do to obtain the next months biostart numbers:

1 Take the three biostart numbers for the first month, and add the number of days in that month, e.g.

JENNIFER	Physical	20 + 31 = 51
	Emotional	1 + 31 = 32
	Intellectual	24 + 31 = 55
MICHAEL	Physical	20 + 30 = 50
	Emotional	28 + 30 = 58
	Intellectual	19 + 30 = 49

SIMON	Physical	6 + 31 = 37
	Emotional	2 + 31 = 33
	Intellectual	26 + 31 = 57

2 If any of the numbers obtained are higher than the maximum number for each biorhythm cycle, subtract the highest number of that cycle until the biostart number falls within a workable range, e.g.

JENNIFER	Physical	51 − 23 − 23 = 5
	Emotional	32 − 28 = 4
	Intellectual	55 − 33 = 22

MICHAEL	Physical	50 − 23 − 23 = 4
	Emotional	58 − 28 − 28 = 2
	Intellectual	49 − 33 = 16

SIMON	Physical	37 − 23 = 14
	Emotional	33 − 28 = 5
	Intellectual	57 − 33 = 24

These figures are now the new biostart values for the next month. Once the initial figure work has been done, this is the only additional process which remains identical for all subsequent months – that is why it is important to record carefully the first set of biostart numbers. This manual method using a calculator was employed for obtaining the initial wave alignment numbers for Jennifer, Michael, and Simon. The specially designed biocharts and bioguides were also used to construct their biorhythmic phasing for the month of interest. Where it is not possible to use colours for the three biorhythms (as in this book), there is another internationally accepted code of representation: for the physical cycle use a continuous line; for the emotional cycle use a dashed line; and finally, for the intellectual cycle use a dotted and dashed line. This is the code that has been used for all illustrated biocharts within this book.

3 WAVES UNDER THE MICROSCOPE

Body charting is a method of keeping track of your hourly, daily, monthly and even yearly body rhythms. If, for example, you note at what periods of the day you feel energetic and at what times you feel drowsy, a pattern will eventually emerge indicating your daily alertness cycle. The purpose of charting your body cycles is simply this: by learning what your rhythmns are, you can plan your activities to best advantage.

BIOFEEDBACK – Marvin Karlins and Lewis M. Andrews

This is where we get down to the nitty-gritty of understanding *how* biorhythm affects us. (After all, there's little point in reading about them if you don't know how you will respond to them.) Our bodies generate a living matrix of countless biological rhythms, ranging from short, rapid ultradian rhythms that complete themselves in minutes, to those that might take a month – the three neo-Fleissian cycles that this book is mainly concerned with. Rhythm science is a valuable tool just waiting to be picked up and used. A tool in your quest for self-discovery (an exciting one) and self-improvement (a gratifying one).

MIND AND BODY – THE INDIVIDUAL

Because of the complexity of our bodies (not to mention the vast intricacies of our minds) it is impossible to make sweeping generalizations about definite effects of biorhythm phasing on the human system. We are, after all, individuals; not robotic automatons. For those of us who have entered adulthood, our school experiences bring back memories that point quite clearly to the fact that each of us possesses varying qualities in the three main biorhythmic spheres of behaviour: physical performance, emotional responsiveness, and intellectual acuity. We will, therefore, react in an idiosyncractic way to our inner energy state.

Medical science has already made considerable advances in the understanding of the mind/body interface. Psychosomatic illness is a reality, with many sufferers of migraine, asthma, peptic ulcers, and cancer often fitting into what is termed a biotype. Certain personality weaknesses and chinks in our emotional armour tend to propagate

47

particular illnesses that are physically very real; these have been germinated in the disharmonious mind. Many philosophical schools of thought point out: 'As a man thinks, so he is'. And, from a purely physical standpoint, we are what we eat. Expanding these principles a little further it can be realized that our thoughts create our attitudes which in turn influence our actions. These encompass all realms of human experience: mental, emotional, and physical. All are profoundly coloured by the very 'timbre' of our body, so diet takes on a significant role in our behaviour whether we are aware of it or not. Many in the Orient employ the system of Yin and Yang for nutrition balance that is embodied in macrobiotics. In the Occident we have come to realize the importance of including vitamins and fibre in our diet, together with a reduction in sugary and fatty food-stuffs as well as white flour products.

There are countless choices that each of us can, and do, make each day. The choices that make us unique, which create our endearing qualities and the not so popular ones. For this reason, predicting how a person will react solely by means of biorhythmic interpretation would be akin to following astrological columns in a daily newspaper. (I remember a colleague of mine who went into business with a friend of his. Both men have the same birth date and therefore share the same biorhythmic highs, lows, and critical days. Their compatibility ratings are one hundred per cent in each rhythm. But, this does not mean that they will react exactly the same way *because* of their biorhythms. They are individuals. Although, I must admit that I wouldn't like to be around when they both pass through a critical day – the chances of an unruffled encounter do seem a bit remote!)

Biorhythm analysis needs to be supplemented by personal observation of your own behaviour, and this closely follows other body charting systems that are used in conjunction wih biofeedback monitoring. This body charting increases your own self-awareness (but not in self-analysis), rather a practical course of self-observation. The period of charting needs to be for at least a month – the longer, the better (and if you can manage it, continuous) – for we are all constantly evolving. Extended charting will be more accurate for you as a whole, and will show up trends in your make-up that cycle at periods of a month and greater. This is by no means unusual. Additionally, you will undoubtedly discover many behaviour trends that are definitely non-rhythmic: perhaps, even, a few personality quirks! This, too, is illuminating, for these trends will most likely be 'reactive', being a response to your environment, rather than innate and cyclic in origin.

One interesting monthly rhythm was found in a research physicist who spent a year in the Antarctic. Under these extreme conditions he was no longer subjected to timing from social cues that most of us experience, for his position was solitary. In addition to this there was an absence of the light/dark cycle that is encountered in more temperate global positions. His simple project in body charting was to record his bed time and rising time. After the year's stay, he handed over the bed time data to two sleep scientists. The results yielded two interesting components: one of them very unusual. Firstly, each day the scientist had been going to bed fifteen to thirty minutes later, and advancing his rising time by an equal amount. (This is quite a common observation in human subjects living in isolation and is known as 'free running'.) After a month, his pattern had shifted so that he slept in the Antartic 'day' and worked throughout the 'night'. Secondly, the strangest fact of all was that every twenty-eight days he somehow managed to resynchronize his rest/activity cycle with his *original* bed time. (It must be stressed that he was totally unaware of this pattern.) Might it be that the twenty-eight day emotional rhythm had some form of timing influence? Did his resetting of sleep time originate with a change in energy associated with emotionally critical days? Or, was there a common body 'clock' for timing the emotional rhythm and his sleep/wake cycle? These questions are not easy to answer, but one thing is certain: each time he reset his bed time, he was encountering *identical* biorhythm conditions in the emotional cycle. This pattern was consistent throughout the twelve month period; a clear case where body charting has thrown up some interesting developments in human periodicity.

SWING HIGH, SWING LOW

Initially, when I have heard people discuss their biocharts for the first time, it is all too frequently interpreted with a mixed bag of emotions. I expect you have heard, or will hear, something like the following: 'Look here! Hey, you've got a double high! You're sure to have a fantastic time ahead, just think of all that boundless energy that you will be able to use . . . there'll be no stopping you!' Or, the prevailing, negative approach turns one of your friends into a Job's comforter: 'Do you really think you will be able to cope with that triple critical *and* those awful lows at the end of the month? What a shame – and all in one month as well. Pity about your holiday then, I *do* hope that it won't be spoiled!'

The fact is that lows are just as useful as highs and both can be utilized equally for your benefit. The whole point about energy fluctuation in these inner rhythms is that you can apply acquired knowledge of your personality to best advantage. Some of us are introverted, others out-and-out extroverts, others lie somewhere in the middle. We might be positive thinkers or face life as one continual struggle. All of us have experienced the enjoyment that we receive when we meet someone who is predominantly positive, and the irritation and heavyness that envelops and taints our perspective when we are in the company of a droning pessimist. It might be, for example, that our energy levels are vastly different. Some of us race through life as if our bodies are secretly E-types in disguise, others are so 'laid back' that one wonders if life is present at all!

High energy individuals will tend to have more accidents and make errors of judgement when passing through positive peak periods as they are, in a sense, overcharged. For them, the best times for steady activity and objective reasoning are during the recharging phases. This lower energy state brings about a balance in their own personality and lifestyle – if it is recognized and employed. Getting in rhythm with your body is the primary aim. Conversely, for those who are perhaps a little more introverted and less inclined to social interaction, the discharging peaks can become very useful springboards into the unknown, undiscovered territories. That 'edge' – crest of the wave – can be all that is needed to provide enough impetus to get those plans moving . . . creative ideas flowing . . . or previously difficult tasks tackled.

It cannot be overstressed that a 'bare bones' interpretation of your biochart without any feedback from body charting and personality awareness will inevitably lead to short-lived successes. So please ignore those who say that: 'HIGHS ARE GOOD, LOWS ARE BAD.' Nothing could be further from the truth.

First of all, I would like to indicate the general trends that are associated with each of the three biorhythms. Within each section I have summarized conditions that you can expect to encounter during mini-critical days (the highest and lowest points of each cycle), high and low periods, and critical days marking the beginning and middle of each cycle. For all summaries excluding the critical day I have given some idea of positive and negative characteristics that can be experienced – this will largely be determined by your own personality. The critical days are excepted from a listing of positive attributes because their effects are generally considered to be unfavourable – although some people do experience 'power surges', especially when passing from negative to positive phases. The effects of the half-periodic day are also noted to be more destabilizing as the body is in the process of entering the recharging span.

After a treatment of each biorhythm you will find a body charting system that you can photocopy and fill out in order to record your daily behavioural patterns. When you have accumulated data for *at least* a month, interpret your biorhythms in the light of this 'biofeedback' and you will be able to make much more accurate predictions about your responses for similar biorhythm conditions in the future. The advantages of this integrated system of body charting and biorhythm profile is that it is *totally specific for you*. Any attempts to fit everybody into similar response states proceeding from identical energy profiles are usually met with confusion for the reasons already given.

With this wealth of information at your fingertips, you will have all that is necessary to further improve those areas of your life that you want to harmonize, and also to expand your capabilities when there is a surplus of vibrant energy stored up and ready for *action*.

THE PHYSICAL RHYTHM

The main modifiers of the physical cycle are one's state of health and degree of fitness. At one end of the physical scale one has the dedicated, professional athlete. At the other, someone who is convalescing in hospital after a major surgical operation. If both people were passing through physical highs it would be absurd to suggest that they will respond with an equal degree of activity due to energy reserves and ability. Similarly, the low periods when the body is building up its resources for the next high will exhibit differing degrees of recharge depending upon the amount of energy released during the previous high. A fit person who has expended little energy prior to the half-periodic day and one who is generally 'run down' (and has also taken on a physically demanding period in the positive phase) will not recharge to the same extent. What you take out of the system has to be put back in again!

The physical cycle is twenty-three days long, and so the positive discharging phase will occupy days one to twelve. The negative recharging phase will then take up the remaining days until the end of the cycle. The positive is associated with an increase in physical energy output – energy consuming activities are, therefore, favourably aspected during this time. When the physical has reached its peak, there is a possibility (with certain types of high energy – hyperergic – individuals) that an accident will occur. This is because all systems are in 'full swing', and one who tends to be very active anyway, might well overstep the mark of their own capabilities without realizing it; then it is too late. For the lower energy type, this is a valuable opportunity to grasp hold of situations that need a little more aggression and 'attack'. All life processes need times of rest and activity; without this design we wouldn't last very long. The lowest point in the recharging phase warrants more attention for someone who is physically inactive as they will tend to find this phase position more trying if too many bodily demands are brought into operation. Studies have shown that infections are more usually caught during the lows – but again, physical health and fitness must be considered. Surgical operations are favoured during the positive span of this rhythm as the body is able to recover from trauma much more readily. Consequently, post operative complications are reduced to a minimum. With these fluctuations in energy, the sensible approach is to monitor your energy output, and simultaneously keep an eye on your food intake in terms of quantity, regularity, and nutrition balance.

When we have an accident, the most likely cause – apart from a lack of concentration – is that we are passing through a period of decreased mind/body coordination. Nerve messages that are continuously passing between our brain and muscles become temporarily ineffective in producing the desired response. The reasons for this are various: we might be physically tired, or worse still, exhausted; we might be suffering from a marked change in body temperature which has upset our metabolism and muscle reaction; we could be experiencing some degree of cramp in our muscles; we might be ill. The ability to respond quickly and effectively in any potentially dangerous situation is obviously important, and it is in this area that research has shown some interesting effects of the twenty-three day cycle.

One of the most exacting professions for accurate mind/body coordination is that of the aircraft pilot. Masses of incoming data has to be rapidly assimilated, and the appropriate decisions made in the form of adjusting numerous dissimilar controls within the pilot's reach. In Missouri Southern State College, Woods and colleagues decided to study pilot performance by testing their physical reaction rate as determined by the rapid sequential pressing of a button – ten times. Five sets of data were taken for each series, and the subjects were tested during various biorhythmic conditions. Their results yielded three levels of physical reaction rate. The average time that the pilots took to press the button ten times was 5.13 seconds. But, when low periods were singled out for analysis, it was discovered that they reacted slower, as it took them 5.32 seconds to perform the same task. With critical days, this dropped still further to an average time of 5.41 seconds.

Willis (who supervised the experiment) concluded that this increase in reaction time that was observed on a pilot's critical day might seem small, but on the flight deck this would be multiplied when having to perform all the various tasks necessary in the throes of a crisis. He strongly advised that this degree of change in physical reaction rates could make the difference between survival for hundreds of people, or a fatal mid-air collision.

The implications for the rest of us are not too difficult to imagine. When we consider the number of motor vehicles that are on today's roads, factors like these should not be ignored. As physically critical days seem to produce the slowest reaction times, drivers would be well counselled to pay attention to these events on their charts – especially if long motorway journeys are being planned. For those of you who work with heavy industrial machinery that needs constant human supervision and maintenance, the physical cycle has its message: watch those criticals and be mindful of the lows! A sluggish response on your part could cost you much more than a severe reprimand from your supervisor. . .

GENERAL TRENDS
Mini-critical (high)
POSITIVE CHARACTERISTICS
Brilliant athletic performance. Fast reflexes. Considerable amount of stamina. High energy output without too much long-term fatigue.

NEGATIVE CHARACTERISTICS
Physical over-exertion. Muscle strain. Attempting too heavy a work load due to over-confidence in physical ability.

Positive phase
POSITIVE CHARACTERISTICS
Sporting activities well aspected, with improved relaxation after games. Good physical coordination and improved mechanical skills. Increased physical drive.

NEGATIVE CHARACTERISTICS
Physically hyperactive and restless. Irritability due to high energy reserves. Inability to tolerate sedentary activities, e.g. sitting down at work.

Critical day
NEGATIVE CHARACTERISTICS
Poor reflex reactions and feeling of lassitude. Very poor coordination. Inability to start the day well. High accident potential. Erratic physical behaviour.

Negative Phase
POSITIVE CHARACTERISTICS
Physically calm and graceful. Ability to tackle less demanding routines

without frustration. Good period for relaxation techniques such as Yoga and T'ai-Chi Ch'uan.

NEGATIVE CHARACTERISTICS
Decreased vivacity and stamina. Quick to tire if engaged in high physical work load. High susceptibility to infections as lowered general resistance.

Mini-critical (low)

POSITIVE CHARACTERISTICS
Good benefits from sustained, non-active relaxation as the body will recharge more efficiently. An appropriate time for less demanding activities, e.g. reading.

NEGATIVE CHARACTERISTICS
Momentary loss of drive, though not so pronounced as the critical day. Depleted stamina with possible muscular fatigue. Accidents possible if precautions are not taken when tired.

THE EMOTIONAL RHYTHM

In the introductory chapter I mentioned some questions that I frequently hear when people ask about biorhythm for the first time. One of them was: 'Do men possess the emotional rhythm?' The answer to this simple question is, simply, yes. This rhythm, because it has a frequency of twenty-eight days and a prevailing influence upon our sensitivity and emotional characteristics, is often confused with, or considered to be, the female menstrual cycle. There is no relationship between these two rhythms. In fact, the menstrual cycle tends to average out at a slightly longer frequency of twenty-nine days – although there is considerable individual latitude in this figure. *(See Chapter 5, 'Nature's Inner Nature, p.80, Women Conceive With a Night Light'.)*

When considering the physical rhythm, two basic modifiers were mentioned: state of health and degree of fitness. With our attention now focused upon the twenty-eight day rhythm, these modifiers are not quite so easily described. Emotionally, we are complex, with attributes that are far less tangible than dense flesh and bone. A well-developed muscle is readily recognized, as are physical injuries, but emotional flaws and inner, untapped resources that surface in a crisis require closer, discerning scrutiny. Our heart-borne emotional 'bodies' are harder to evaluate for they cannot be seen with the naked eye, but our responses and behaviour can be: the end-products of our essential nature. Of all the rhythms, this one seems to present the greatest degree of difficulty to a student of biorhythm because they often do not know what they are *feeling*. This comes within the realm of self-analysis; not an easy task. But, behaviour observation is much more definable and equally (if not more) valuable in determining your emotional 'tone'. The interpretation of nebulous feelings is prone to greater error of judgement than factual recording of your interaction with people, 'loner' episodes, or creative adventures. These initial problems of assessment can be overcome by a thorough body charting in order for you to 'fine tune' your cyclical awareness plus evaluate those traits of a more incidental nature. In time you will be able to recognize those aspects of your personality that are directly affected by this rhythm. However, do not expect to be able to achieve this overnight, remember that this cycle is nearly one month in duration.

The first fourteen days of this cycle are associated with outward, creative, self-assertive qualities. As our emotional state plays a valuable part in day-to-day living, it follows that this rhythm has a marked influence on our conduct and manner. If we 'feel' that a task can be tackled – approaching the boss for a rise, confronting a partner with grievances that have been hitherto swept to one side, or even taking the 'plunge' and proposing to a sweetheart – it will be, even during adverse conditions in the other cycles. People tend to be more communicative during the emotional highs, enjoy the stimulation of working in teams, and also take to new, enterprising situations because their whole system is 'geared up' and naturally spontaneous. During the peak, those that are more excitable would do well to carefully consider any decisions that have been made on impulse: that 'high' might create an over-confident approach that terminates in a humiliating withdrawal. For those that are more placid, the highs will coax them out of their shells, revealing to others those inner qualities that they possess which *need* to be developed and used. So, if you fit into this category, utilize these periods to the full – you will surprise yourself with fresh accomplishments.

From days fifteen to twenty-eight of the cycle, the rhythm is in the negative phase; a time of decreased emotional vivacity, and a time to recharge the batteries of your 'heart'. For those who are more extroverted, this phase will hail a period of calmness, when a more tranquil state of mind will readily ensue if you set some time aside. This is nature's way of bringing about a balance: there is a time for creation and a time for receptivity. During lower-key phases, the mind will be more amenable to mulling over ideas that will later be brought into operation when conditions allow. Additionally, a degree of emotional detachment will often bring a 'breath of fresh air' as situations can be regarded from a more objective viewpoint, rather than the thinker being too involved and caught up emotionally. For others, the recharging aspect of this wave will dictate a concerted effort to overcome any tendency towards self-centred thinking or unhealthy, prolonged isolation should these traits be dominant in the personality. Irritability and overcritical attitudes have also been observed more frequently during this phase. However, emotional lows do not necessarily generate negative thinking patterns: some will be more susceptible to sombre outlooks than others by reason of their disposition.

GENERAL TRENDS
Mini-critical (high)

POSITIVE CHARACTERISTICS
You will tend to be more extroverted and communicative with friends. Surges of creative gusto and feelings of well-being and inner assurance. Ability to tackle problems in a positive fashion.

NEGATIVE CHARACTERISTICS
Over-confident in ideas and plans. Inability to keep confidence, tactless verbal behaviour. Possibility of becoming too involved in your own dreams rather than seeing the realities of life around you.

Positive Phase

POSITIVE CHARACTERISTICS
You feel content with yourself. Artistic designs often come to fruition now. If you have any leadership qualities, now is the time to put them into action. A period of innovation.

NEGATIVE CHARACTERISTICS
High self-interest and little concern with the welfare of others. Tendency to be too impulsive with a lack of solid planning. 'Flash in the pan' schemes might start here!

Critical Day

NEGATIVE CHARACTERISTICS
Feelings of instability and insecurity. Unsure of affections from loved ones. Loss of confidence and inability to visualize your plans coming to fruition. Momentary loss of direction and purpose in life.

Negative Phase
POSITIVE CHARACTERISTICS
A good time for *cooperating* with group aims and getting involved in social activities. More reflective states of mind will ensue. Ability to emotionally detach yourself from any difficult situation that needs a rapid, logical decision.

NEGATIVE CHARACTERISTICS
Withdrawal might be the order of the day. Dislike of communication –
'loner' feelings. Critical attitudes to colleagues at work. Irritability
appears with any sudden change of routine.

Mini-critical (low)

POSITIVE CHARACTERISTICS
A good time for a cool, level-headed assessment of your aims and
direction in life. Your head rules your heart. Any isolated activities will
be taken with ease.

NEGATIVE CHARACTERISTICS
Possibility of feeling depressed and unfulfilled as everything looks
'black'. Difficulty in coping with highly social occasions, e.g. parties,
conferences, lectures. Lack of momentum to put your plans into
practice.

THE INTELLECTUAL RHYTHM

Finally, we consider the longest rhythm of all; intelligence is one of the
main modifiers of this cycle. Considering the Law of Initial Value,
envisaged by Dr Joseph Wilder of New York University, it follows that
each of us will respond differently during intellectual highs, lows, and
critical days. Our mental capacity, memory ability, degree of education,
and day-to-day thought processes will modify the way in which we live.
We might be slow, ponderous thinkers that steadily plod on to the goal
in sight, letting nothing distract us from the desired aim. On the other
hand, we might exhibit erratic flashes of quick thinking that are
interspersed with periods of mental quiescence. Our minds could be
balanced between the two extremes, enjoying the best attributes of a
steady stream of 'low level' thought plus the occasional brilliant idea or
'brain wave'. Some of us are divergent thinkers, often perceiving reality
from a bird's eye view. Many of a philosophical bent will reveal this
type of mental processing, whilst the scientific specialist will often be a
success due to his convergent powers of reasoning, focusing all
attention on a single problem. Each of us tackles life's situations in
quite different ways, for our background, heredity and training have
developed in us certain mental patterns that are as definite as a tapestry.
The fabric of our thoughts can be the key to success by being
constructive, or hinder our progress through being detrimental to our
well-being.

The positive phase of the intellectual cycle lasts for sixteen-and-a-half
days, and is, like the other biorhythms, balanced by a negative phase of
an equal amount. Considering first the positive phase, the following
attributes are associated with this biorhythm condition:

As this cycle discharges, thought patterns are usually more energetic and spontaneous. Decision making in this phase will tend to be more rapid, with problems of a demanding nature offering less resistance as the mind will have greater powers of concentration, perseverance, and skills of deductive, logical reasoning. For the student, this phase signals times of increased learning and is more suited to the acquirement of new knowledge rather than reviewing retained information. Your ability to assimilate facts and assorted data are usually at a peak during these times. Any area of life that demands some sort of choice based on accurate assessment of prevailing conditions – rather than an intuitive hunch – will be valuably enhanced by this phasing, whether it be planning your financial affairs for the next few months, deciding how best to redecorate the house within a restricted budget, or redesigning a recipe that has spelt rock-hard disaster every time the microwave door has been opened!

As so much of our lives depend upon making sound decisions, this cycle warrants just as much attention as the physical and emotional cycles, even though initial research has suggested that this longest cycle affects our behaviour the least. With this ability to absorb information more readily, the positive phase is also associated with increased verbal communicative skills. (If there have been any emotional problems that you have had particular difficulty in expressing to someone close to your heart, this could well be the right time to air your feelings.) Meetings involving groups of people where complex issues are being debated are also favourably aspected, provided that you do not react during the peak by being too assertive and arrogant in your own opinions. Another aspect of this phase is that creative pursuits are often brought to fruition because any ideas that have had time to develop (perhaps in an emotional high) now gel with the added momentum of an alert mind.

Having passed through a half-periodic day, the intellectual cycle will enter the recharging phase. This is generally linked with decreased mental energy – but not intelligence – this transition does not mean that we mutate from budding Einsteins to hapless cabbages! This sort of crude construction on biorhythms will no doubt be a great source of amusement to some, but they will have to search long and hard to find it within this book (or, for that matter, within any other published material on biorhythm).

During this lessened energy phase, there are many tasks that will be found more agreeable than during a mentally vibrant condition. It is prudent to consider this balance of mental activities, for we cannot be operating at full swing all of the time. For those who race around constantly devising new plans and ideas, this phase pleads rest to the grey matter. It is only when we pause from our activities that we are able to obtain some degree of perspective in our daily affairs. Often, a few minutes spent reviewing our pursuits, interests, and natural talents will be amply rewarded by gaining a sense of direction and purpose in life. This phase does not mean that all intellectual pursuits will grind to a halt – as if someone has just pulled the plug from your mental power pack! It merely depicts times of decreased drive and acuity – more apparent in some than others, and depending on profession, how influential – the professional footballer need not pay so much attention to this cycle as his accountant. It is not a bad idea to make a point of changing a few routines so that your mind will be stimulated into thinking about life in a new way. Change your habits when you come home from work, and also make full use of the weekend, exploring original possibilities and avenues for enjoyment and relaxation. Allow the suggestions of friends to help you along the way. This fresh approach will alleviate any feelings of the humdrum routines of life and any mental lassitude if present. It also follows that if you are in a working mood and have any repetitive tasks lined up – and have felt too innovative at other times to be bothered with them – these can now be approached in a routine and steady manner that will come naturally to you at these times.

GENERAL TRENDS
Mini-critical (high)

POSITIVE CHARACTERISTICS
Your intellectual perceptions are razor sharp. Judgement skills are fully enhanced. Excellent memory retention/recall. A good time for expressing those well organized objectives.

NEGATIVE CHARACTERISTICS
You quickly pour 'cold water' on other people's plans. You are too strident with your own aims and impulsive. Have you overrated your ability to carry a mentally demanding task through to the end?

Positive phase

POSITIVE CHARACTERISTICS
Good organizational skills – time to plan that weekend away with friends. Sound, objective decision making can be called upon when needed, so a good time for financial planning.

NEGATIVE CHARACTERISTICS
Inability to stick with a course of action; you're a bit 'grasshopper minded'. Your behaviour might appear a bit aimless if you don't channel your ideas a bit more.

Critical day

NEGATIVE CHARACTERISTICS
Loss of concentration. Mental dullness and weariness are more evident. You are easily distracted from the task in hand. Don't make any important decisions today.

Negative phase

POSITIVE CHARACTERISTICS
Leave the innovations alone and concentrate on routine tasks that you are familiar with. A favourable time to review your work. Catch up on all that light reading. Time to 'stand back' and consider your next move.

NEGATIVE CHARACTERISTICS
Research involving intellectual stamina will be a struggle; not a time for pressing ahead with new ideas. Learning new skills will be less favourably aspected.

Mini-critical (low)

POSITIVE CHARACTERISTICS
A good time for relaxation and a change of routine to stimulate your mind and revitalize your attitudes. Do something out of the ordinary after work today. Break the pattern.

NEGATIVE CHARACTERISTICS
You will be less independent with the realization that good advice from others should not be ignored. Be extra vigilant against repercussions if your plans have gone awry. Delegate responsibility where possible.

BODY CHARTING

The following questions have been designed to locate areas of your behaviour that are constantly fluctuating. In total there are thirty questions and they can easily be answered in about ten to fifteen minutes, with practice you will speed up. This is all the time that you need to put aside each day in order to build a more in-depth analysis of your cyclical behaviour patterns and those that are evidently non-rhythmic. At the back of this book you will find a chart that you can photocopy, with spaces left for your name, date, and answer. After a sufficient amount of time has been spent on body feedback, you will be able to view the results in perspective against your biorhythm profile.

By putting the two systems together, more specific trends will be detected in the physical, emotional, and intellectual cycles. No two people will modify their biorhythm phasing in quite the same way, but, as indicated, the general patterning still stands. Some might find that unusual 'power surge' on an intellectually critical day, others with an extroverted personality will feel calm during emotional lows, rather than a mood of pronounced withdrawal from their friends. For sporting types, physical highs might reveal brilliant performance (as for Mark Spitz when he won seven gold medals in the 1972 Olympics – both emotional and physical rhythms were high). Others, who are equally active, could be overcharged by these identical conditions and overstep their capabilities: for instance, an embarrassed moment on the squash court due to a strained muscle. The fun begins here, so keep up the records and start to improve your life through intimate knowledge of your life patterns. There will be many surprises along the way.

PHYSICAL

1 What was your bed time?
2 What was your rising time?
3 Have you had any naps during the day (time and duration)?
4 What has been your degree of physical energy during the last twenty-four hours (inactive, average, very active)?
5 What is your weight (take at the same time each day)?
6 Type and duration of physical exercise during the day (jogging, football, tennis etc.)?
7 Type and duration of sexual activity during the last twenty-four hours (none, masturbation, intercourse)?
8 Physical aptitude during the day (clumsy, normal, well coordinated)?
9 How would you describe your state of health today (ill, normal, very fit)?
10 How have you eaten during the last twenty-four hours (poorly, normal, gluttonous)?

EMOTIONAL

11 How communicative have you been during the last twenty-four hours (taciturn, normal, over-talkative)?
12 Have you enjoyed company or preferred to be on your own during the day?
13 How creative were you today (verbal, written, or artistic)?

14 Have you *felt* understood or misrepresented at work, home or with friends today?
15 Have you made any new friends today?
16 Have you had any serious arguments today?
17 Generally, have you been relaxed, or under stress over the last twenty-four hours?
18 Have you had feelings of rejection or approval *towards* your closest friends today?
19 What has been your general mood throughout the day?
20 How would you describe your outlook on the day's activities (negative, mixed, positive)?

INTELLECTUAL

22 How well have you been able to concentrate on your work during the past twenty-four hours (poor, average, good)?
22 Have you expressed yourself well today, or had great difficulty in communicating your ideas?
23 Have you been able to solve quickly the problems of the day, or have they taken longer than usual, or average?
24 Has your day been marked by spells of forgetfulness?
25 Have you had to repeat work through careless mistakes?
26 Have you found it relatively easy to retain the day's information (written, verbal, or visual)?
27 When decision making, have you been able to arrive at a conclusion quickly, with difficulty, or in average time?
28 Have you been over-confident in your ideas today, as colleagues thought them unsound?
29 Today, have you preferred to revise past plans and information or forge ahead and construct new objectives?
30 On the whole, would you describe your mental processes over the last twenty-four hours as ponderous, adequate, or very responsive?

*   *   *

Remember that you respond to all three biorhythms at the same time, so keep this in mind when interpreting your biochart for improved forward planning.

4 ARE YOU OUT OF STEP?

There is a new conception of man's capacities forming within the scientific community. During the twentieth century, academic science has radically underestimated man – his sensitivity to subtle sources of energy, his capacity for love, understanding and transcendence, his self-control.

DR JAMES FADIMAN – Stanford University

What makes some people 'click' within a short time of meeting one another, whilst others seem destined continually to rub each other up the wrong way? Why is it that you have to work really hard to establish a good business relationship with someone who you see each and every day? Yet, you have some friends that you look up only once or twice a year, but it seems only yesterday: you both pick up the threads of your friendship with ease, almost as if you were never apart. In this case, absense does make the heart grow fonder. There are some people that you catch sight of frequently and yet you know that they will never be anything more than just plain acquaintances. Others will occupy that special place within your psyche as the closest of friends. Similar personalities and common interests don't always seem to provide all the necessary ingredients for a successful relationship, and those who you thought of as unlikely companions because of their 'alien' life-style, soon develop a rapport with you that is both surprising and stimulating.

YOUR ENERGY PROFILE

Perhaps you have often been mystified about the nature of friendship. Biorhythm compatibility studies reveal that each of us has an energy profile that is 'activated' when we come into contact with other people. Sometimes this profile will key-in to other people's dynamic configuration, for they are similar in nature. With other individuals, a clash occurs because your vitality trends will always be antagonistic to that person. This profile can be determined by evaluating how well your physical, emotional, and intellectual rhythms are aligned with the corresponding rhythms in the person of your choice.

When we ponder those relationships that work really well, it becomes apparent that similar aims and goals help to cement that invisible bond that exists between two like minds. Common purpose is enormously enhanced if each individual is able to follow or match his or her partner's physical energy, creative ideas, or intellectual drive. This ideal situation in all three rhythms is something that most of us have not encountered, but this is precisely where the employment of your own energy profile becomes a valuable asset.

With team work, it is important that the group is able to work together well; the stability of the whole is dependent upon a cohesion of minds and energies so that collective achievements are far greater than any individuals accomplishments. This is not only brought about by a dynamic mix of personalities, but also by the energy of the group flowing into constructive forms that can be easily channelled. There is little point in one person going through an energetic, positive thought process if everybody else responds by throwing 'cold water' on each sentence that is uttered. It is in this area that compatibility analysis can reveal weak links in the group; appropriate action can then be taken.

COUNTING THE DAYS

In chapter two you learnt how to count up the number of days from your birthday to the month of analysis, and then process this number in order to discover your biorhythm phasing. The principle is very similar when we want to find out the compatibility rating between two people. In this situation we have to add up the number of days *between* the former and latter birth dates for the two individuals concerned. It follows that your energy profile will be different with each person and that it remains fixed throughout your life. This profile is usually expressed in terms of three percentages that indicate the degree of alignment for each pair of cycles. The degree of alignment will always be the same for two people because their corresponding cycles will continually be exhibiting the same amount of phase difference to each other. (Although the individual rhythm will go high, critical, and low, they will always fall behind, or be ahead of, each other by that calculated amount.)

THE TWO EXTREMES AND IN BETWEEN

Initially, these alignment percentages might seem a little difficult to interpret, but the concept becomes a lot clearer if one considers the two extremes in a relationship: totally in phase and totally out of phase. It then becomes a simple matter to visualize the prevailing state for values that fall within this range. For instance, if someone's energy profile indicates that they are zero per cent compatible with another person in their emotional cycle, it means that they are totally out of phase with each other. When one experiences an emotional high, the other will be passing through an emotional low. The reverse would also be true. Additionally, this zero condition means that their critical days will align. The same can be said for the other two rhythms: zero to four per cent means opposed phasing and shared critical days. If two people investigate their compatibility rating and find that their intellectual rhythms are rated at one hundred per cent, they both experience highs at the same time, lows are equally matched, and critical days are coincident. Or, to put it another way, their mental activity cycles will pass through the same conditions at the same time. All intermediate values between these two polar aspects reveal a gradation of correspondence or opposition between the cycles concerned. The various combinations of energy phasing between two individuals provide us with a countless number of cyclic environments that have the ability to govern a large proportion of our interactions within personal relationships. Biorhythm compatibility analysis is a means of understanding these conditions so that we can look objectively at our behavioural patterns from this 'energetic' standpoint.

WORKING OUT YOUR COMPATIBILITY PROFILE

To work out a couple of examples of how you might calculate your biorhythm compatibility with another person, let's consider Jennifer Brown, Michael Hanson, and Simon Watkins again. Their respective birth dates are as follows: 8 February, 1966; 14 October, 1964; 8 August, 1957. The following calculations will evaluate Michael's compatibility with Jennifer, and Simon's compatibility with Michael.

1   If the two birth dates are more than one year apart, count up the number of years difference and multiply this number by 365 to obtain the number of days, e.g.

MICHAEL/JENNIFER   14 October, 1964 to 13 October, 1965 = 1 year. 1 × 365 = 365 days.

SIMON/MICHAEL   8 August, 1957 to 7 August, 1964 = 7 years. 7 × 365 = 2,555 days.

2   Count up the number of days from the last date to the end of that month, e.g.

MICHAEL/JENNIFER   13 October, 1965 to 31 October, 1965 = 18 days.

65

SIMON/MICHAEL   7 August, 1964 to 31 August, 1964 = 24 days

3   Count up all the days for all subsequent months up to the month *preceding* the latter birth date. Refer to the *Table of Days in each Month* at the back of the book, e.g.

MICHAEL/JENNIFER   November through to January. $30 + 31 + 31 =$ 92 days.

SIMON/MICHAEL   September. 30 days.

4   Count up all the days in the month of the latter birth date up to the date *before* the birth date. (If the birth date is the first of the month then make no additions in this step), e.g.

MICHAEL/JENNIFER   1 February, 1966 to 7 February, 1966 = 7 days

SIMON/MICHAEL   1 October, 1964 to 13 October, 1964 = 13 days.

5   Count up all the extra days due to leap years that have occurred between the two birth dates. Refer to the *Table of Leap Years* at the back of the book, e.g.

MICHAEL/JENNIFER   14 October, 1964 to 8 February, 1966 = 0 leap days

SIMON/MICHAEL   8 August, 1957 to 14 October, 1964 = 2 leap days

6   Add up all the days calculated in steps (1) – if applicable, (2), (3), (4), and (5), e.g.

MICHAEL/JENNIFER $365 + 18 + 92 + 7 + 0 = 482$ days
SIMON/MICHAEL $2555 + 24 + 30 + 13 + 2 = 2,624$ days

7   The number obtained in (6) is the total number of days difference between the earlier birth date and the latter birth date. This number must then be divided by the day lengths of each biorhythm cycle, e.g.

MICHAEL/JENNIFER   Physical $482/23 = 20.956521$
                   Emotional $482/28 = 17.214285$
                   Intellectual $482/33 = 14.60606$

SIMON/MICHAEL   Physical $2,624/23 = 114.08695$
                   Emotional $2,624/28 = 93.714285$
                   Intellectual $2,624/33 = 79.515151$

Like the previous calculations in chapter three, you will usually obtain several numbers after the decimal point, but these can be ignored. *Reduce* the resultant number to the nearest whole number if necessary.

8 Multiply the number obtained in (7) by the respective number of days that it takes each cycle to repeat itself, e.g.

MICHAEL/JENNIFER Physical $20 \times 23 = 460$ days
                  Emotional $17 \times 28 = 476$ days
                  Intellectual $14 \times 33 = 462$ days

SIMON/MICHAEL   Physical $114 \times 23 = 2,622$ days
                  Emotional $93 \times 28 = 2,604$ days
                  Intellectual $79 \times 33 = 2,607$ days

9 Next, subtract the number of days obtained in (8) from the number of days obtained in (6), e.g.

MICHAEL/JENNIFER Physical $482 - 460 = 22$ days
                  Emotional $482 - 476 = 6$ days
                  Intellectual $482 - 462 = 20$ days

SIMON/MICHAEL   Physical $2,624 - 2,622 = 2$ days
                  Emotional $2,624 - 2,604 = 20$ days
                  Intellectual $2,624 - 2,607 = 17$ days

10 These final figures indicate the difference in days between each respective pair of cycles, and can be used to locate the degree of alignment in the *Table of Biorhythmn Compatibility Percentages* at the back of the book. Locate the figure obtained in (9) in the first column of the table (Days between each cycle). Follow the row across to the correct column for the cycle concerned and find the compatibility percentage, e.g.

MICHAEL/JENNIFER Physical   22 days equals 91%
                  Emotional   6 days equals 57%
                  Intellectual   20 days equals 21%

SIMON/MICHAEL   Physical   2 days equals 83%
                  Emotional   20 days equals 43%
                  Intellectual   17 days equals 3%

At the back of the book you will find a blank table that you can photocopy and fill out to help you through the process. Each step that you have just gone through is similarly numbered on the table. The following illustration shows the table filled out for Michael Hanson and Jennifer Brown.

The results in step (10) reveal the energy profiles for Michael with Jennifer, and Simon with Michael. Jennifer's compatibility profile with Michael will be the same as Michael's with Jennifer. The same is true for all profiles as the results are dependent upon the number of days between the former and latter birth dates. These percentages can then

PERS. 1	M. HANSON		DOB	14.10.64	
PERS. 2	J. BROWN		DOB	8.2.66	
1	YEARS × 365		=	365	
2	DAYS TO END MONTH		=	18	
3	DAYS IN MONTHS		=	92	
4	DAYS IN BIRTH MONTH		=	7	
5	LEAP DAYS		=	0	
6	TOTAL OF (1),(2),(3),(4),(5)		=	482	
7	(6) / 23	(P)	=	20.956521	
	(6) / 28	(E)	=	17.214285	
	(6) / 33	(I)	=	14.60606	
8	(7P) × 23	(P)	=	460	
	(7E) × 28	(E)	=	476	
	(7I) × 33	(I)	=	462	
9	(6) − (8P)	(P)	=	22	
	(6) − (8E)	(E)	=	6	
	(6) − (8I)	(I)	=	20	
10	(9P)	(P)	=	91	%
	(9E)	(E)	=	57	%
	(9I)	(I)	=	21	%

be classified according to various categories. It is pointless making too fine a distinction between results that are only a couple of integers apart when comparing two separate profiles; for this reason compatibility values are subdivided into four main groups, with two smaller groups at either end of the scale for the special conditions of totally in phase and totally out of phase. The groups are as follows:

Reversed	0-4%
Poor	5-25%
Fair	26-50%
Good	51-75%
Excellent	76-94%
Aligned	95-100%

From these groups it can be observed that the energy profiles for Michael with Jennifer, and Simon with Michael are really quite similar. Both contain excellent physical compatibilities, good and fair emotional compatibilities, and poor intellectual compatibilities. A summary of their profiles will indicate the general conditions that they can expect if they see each other regularly. (In all these calculations, individual birth times are not taken into consideration. In many profiles it may be possible for two people to alter their alignments by anything up to one full day in either direction due to birth time. All percentages in the *Table of Biorhythm Compatibility Percentages* are based on the time of birth as being midday.)

MICHAEL AND JENNIFER'S PROFILE

The physical compatibility is ninety-one per cent which means that their cycles are only one day out of step; their highs, lows, and critical days will be separated by only twenty-four hours. This is a very high compatibility, so it means that they will be almost totally matched in physical energy trends. Similar pursuits can be entertained and enjoyed, assuming that they both possess a reasonable degree of fitness. Although highs will be characterized by periods of activity and action, lows and critical days will need attention as they will both be going through the same conditions. Emotional compatibility is good, and it is often quite acceptable to have rhythms that show this degree of correspondence. In this type of alignment, their emotional energies will often be found in the same phase, but at differing levels. This will lead to a healthy balance of emotional outlook through moderate contrast: neither too closely followed or too oppositional in nature. With this middle-of-the-road result, many relationships will work really well that might suffer if the compatibility was at a higher level, where too similar conditions might propagate irritation. Their intellectual value is rather low, and falls within the poor sector. At twenty-one per cent their rhythms will be in a similar phase for about one quarter of each calendar month. If their relationship is strongly governed by mental activities, and they are both individuals that feel a great need to communicate their ideas and feelings to each other, then this result will indicate problems on the horizon. On the other hand, if they are both independent, and have a tendency towards being taciturn, this rhythm will have less influence on the overall maintenance of the relationship. With this poorly aspected alignment, organizational efforts might be difficult to coordinate between the two of them.

SIMON AND MICHAEL'S PROFILE

For the physical rhythm, similar conditions exist as in Michael and Jennifer's profile: Simon and Michael's rhythms are only two days apart. In this male to male relationship, team sports are very favourably aspected, and other joint physical activities, such as jogging, because physical strength will usually be more closely matched than in a male to female situation. There will be less problems when encountering critical days as they are forty-eight hours apart: although a narrow margin, this does give more of a breathing space than one day, where unstable conditions will probably overlap. However, lows will overlap, so particular attention must be paid to these periods, especially if strenuous activity is planned: neither Simon or Michael will be in the best of form. Emotional compatibility falls within the fair range, so on the whole, their twenty-eight day cycles will be in opposite phases. Depending on the type of relationship studied, this can have quite a noticeable effect. In this instance, a friendship is being analysed, and therefore Simon and Michael will spend less time in each other's company than a married couple, or two people who work within close proximity to each other. Differences in social sensitivity or emotional disposition will, as a result, be less inclined to have much bearing on the relationship, unless stressful conditions of the day bring any underlying inclinations to the fore. Coming finally to the intellectual cycles, as in Michael and Jennifer's physical rhythms, Simon and Michael's intellectual rhythms exhibit critical days that are only one day apart. To be more precise, they are twelve hours apart as the half cycle in this longest rhythm is 16.5 days. It is not possible to obtain a zero per cent compatibility in the intellectual cycles, so this three per cent is the lowest score. Simon and Michael are totally out of phase with each other. When one is high, the other is low. Successive 16.5-day periods will find their situations reversed. Critical days will be coincident. If their companionship is founded upon sporting activities, this condition will have little influence on the friendship. If, on the other hand, they enjoy competing on an intellectual level, such as playing chess or bridge, a few easy wins will often be due to totally opposite phasing. As for those games that end up with both parties wishing they had never started, joint critical days will be the culprit!

To help you focus your attention on the results of a compatibility analysis, I have treated each percentage group separately and given some guidelines about the energy conditions that you might encounter. It is also important to take into account each individual's state of health, emotional make-up, and mental qualities. The relevant sections in chapter three cover these behaviour modifiers more fully.

PHYSICAL COMPATIBILITY 0-4%

Your physical rhythm is the reverse of your partner's physical rhythm. When you are feeling physically energetic, your partner will be going through his/her lowest vitality point. Conversely, when you feel like resting, your companion will be experiencing the most physically active period. This reversed condition can create relationship difficulties if you do not make any allowance for differences in stamina. You will also share the same critical days. During these unsettled periods you will tend to be more accident prone, as your reflexes will be slower; other body responses will be well below par. These episodes occur every 11.5 days, so both of you must keep a diary of your critical moments and take extra precautions against physically stressed situations that might be potentiated by this condition.

PHYSICAL COMPATIBILITY 5-25%

Your compatibility is poor. This means that for most of the time, your physical cycle will be opposed to your partner's equivalent cycle. When you experience periods of fatigue or high energy, it is more likely that your partner will be encountering the opposite. As a result of this poor compatibility, you will have to show a fair amount of understanding when pursuing joint physical activities such as team sports, disco-dancing or even jogging. The best approach is to try and compromise when the going gets tough. At times your partner might want to take forty winks, whilst you have already packed all the swimming gear! (Or vice versa for that matter.) Understanding the reasons for these differences in physical drive and stamina will go a long way to improving this aspect of your relationship if it plays an important role in your lives.

PHYSICAL COMPATIBILITY 26-50%

Your physical compatibility is fair. On balance, the physical biorhythm will not be in harmony with your partner's corresponding rhythm. This physical cycle affects your vitality, stamina, and muscle power, so pursuits involving a lot of activity come under its influence. If you want to go for a long walk in the countryside, and your soul mate prefers to stay in, your biorhythms are probably opposed. However, for some of the time, your energy levels will be in a similar phase. Take advantage of these shorter spells of similar energy when they arise. In these periods, activities that require an equal amount of stamina from the two of you can be entertained. This low compatibility can be minimised if both of you try to cooperate with each other when differences in physical stamina occur.

PHYSICAL COMPATIBILITY 51-75%

Your compatibility is good. When compared, the physical cycle will tend to be at a similar energy point to that of your partner's corresponding cycle. Team sports such as tennis, squash, and football will often be mutually enjoyed. This physical rhythm controls your staying power, reflexes, and bodily strength, so when you seem to be experiencing one of those 'off' days, your partner, on balance, will probably be feeling the same. Generally, there shouldn't be too many clashes of physical drive – if this does occur, then hold on! As these rhythms are quite well matched, opposed periods will be shorter than those of a similar biorhythm position. During these lesser spells of disharmony, try to meet half-way, knowing that the body needs its periods of rest and activity in order that health may be maintained.

PHYSICAL COMPATIBILITY 76-94%

Your compatibility is very good. For most of the time, your physical cycle will be at a similar potency level to your friend's. As a result of this, shared physical activity will usually find both of you with an equal degree of stamina and muscle power. Because of this good match in physical energy trends, you will probably tackle tasks together without much problem. This high compatibility can also have its drawbacks: when you are feeling fatigued, do not expect too much of your partner as they will be in a similar biorhythm position. During high periods, it will be easy for both of you to tackle too much: the trend being for a reciprocal increase in physical drive. So, be more watchful than usual during these peak intervals.

PHYSICAL COMPATIBILITY 95-100%

Your physical rhythms are exactly in phase. This cycle modifies agility, energy reserves, coordination, and stamina. As your rhythms are matched, it follows that you will share the same highs and lows, so mutual involvement in physical pursuits should present little difficulty. However, this total alignment can have its limitations: you will tend to accentuate each other's highs and lows, resulting in too ambitious, or leisurely a material objective. When you pass through a resting period, don't be surprised to find that your partner is in a similar position. Hence, look to other friends for physical support when you are flagging. Physically critical days will also be coincident – these occur every 11.5 days – so be on the alert as both of you will be more accident prone than during other biorhythm conditions.

EMOTIONAL COMPATIBILITY 0-4%

Your emotional rhythms are totally out of phase. This cycle controls creative drives, inspirational ideas, moods, and more importantly, your sensitivity to friends and acquaintances. When you are passing through an emotional peak, your partner will be at the lowest point of this rhythm. As a result, conflicts might often occur: when you harbour a more introspective mood, your companion will most likely be emotionally outgoing and active. A more constructive approach to this polarized state is that you will be promisingly situated to balance each other's downs, and to temper those too exuberant highs. You will also share the same critical days which occur every fourteen days (the same day of the week that you were born). During these twenty-four hour periods, undue sensitivity can lead to frayed tempers, so take special note of these mercurial events.

EMOTIONAL COMPATIBILITY 5-25%

Your emotional compatibility is poor. This means that for most of the time, your emotional rhythm will be in a different phase to your partner's corresponding rhythm. This cycle modulates your outward sensitivity, inner feelings, moods, and intuitions. With this poor compatibility, it is likely that you could experience temperament clashes with your partner: when you are feeling extrovert and chatty, he might prefer a quiet day without any interruptions. As your biorhythm profile changes from day to day, the reverse could also be true – you might feel like being alone for a while. Knowing this, it will be easier to balance each other's moods when the need arises. The old adage 'forewarned is forearmed' certainly helps if we want to build up a stronger relationship.

EMOTIONAL COMPATIBILITY 26-50%

Your compatibility is fair. On balance, your emotional rhythms will be in differing phases. This cycle controls your moods, creative ability, and receptivity to friends. As the compatibility is low, you might often encounter a dissimilarity of affections and emotional energy in the relationship: when you feel full of creative drive and gusto, your soul mate might be in a more reflective state of mind. The situation could be reversed: you might be the lone wolf for a while, but your companion seeks lots of company. In view of this trend, try to balance each other's emotional highs and lows. For some of the time, your emotional energy will be in the same phase as your partner's – use these times constructively to strengthen the relationship.

73

EMOTIONAL COMPATIBILITY 51-75%

Your compatibility is good. On balance, each month will find your emotional cycle in a similar phase to your partner's equivalent cycle. This rhythm governs your creative drive, sensibilities, and attitudes to colleagues and friends around you. This similar phasing means that you will often follow each other's emotional energy; use these times jointly to pursue activities influenced by this rhythm. But, bear in mind that occasional clashes will occur as these two cycles are not totally in step. These opposed periods can be used constructively as well: try to balance each other's feelings and attitudes when differences arise. When your partner seems withdrawn, you might be in the right frame of mind to coax him out of that shell. On the whole, the two of you should be fairly well matched in this twenty-eight day rhythm.

EMOTIONAL COMPATIBILITY 76-94%

Your compatibility is very good: your emotional cycles are closely aligned. This rhythm affects your reactions and responses to your immediate environment in the emotional realm. As the compatibility is high, you will tend to encounter highs and lows at similar periods to your companion. Watch out for those days when you might be feeling negative; your friend will probably be sitting in the same boat, trying to weather the same storms. This trend is conducive to a natural understanding of each other's mood changes and emotional energies, but extreme states might prove a little difficult to balance as both of you are in a similar biorhythm position. Therefore, use your intuitive understanding of each other to tackle difficulties when they arise.

EMOTIONAL COMPATIBILITY 95-100%

Your emotional rhythm is totally aligned with your partner's corresponding rhythm. This cycle guides your creative potentiality, heartfelt impulses, and spontaneity. As these cycles are in phase, you can expect to share life's ups and downs, so emotional responses should be easily comprehended. Make the most of your high periods, when you will be able to achieve much more as a team than either of you can achieve alone. Watch out for those low periods too, as a negative state of mind on your part could be readily absorbed by your friend. You will also share the same critical days, so note these unstable occasions that occur every fourteen days (the same day of the week that you were born). Concentrate on communicating to each other during highs as divergent opinions are best dealt with at these times.

INTELLECTUAL COMPATIBILITY 0-4%

Your intellectual rhythm is opposed to your friend's intellectual rhythm. This means that when you encounter a low phase, your friend will be at the highest point, and vice versa. This cycle controls your thought processes such as judgement acuity, memory retention/recall, and the ability to express your feelings to those around you in an articulate manner. As your intellectual rhythms are always opposed, you will probably encounter some communication breakdown at times: when your mind is just 'idling' – ready for the next high – your partner will be at his or her most active state, trying to impress upon you all of those fast-flowing ideas and plans. Try to realize this aspect on the mental plane as it will help to alleviate problems when they materialize; all of us need regular periods of mental rest and activity.

INTELLECTUAL COMPATIBILITY 5-25%

Your intellectual compatibility is poor. As a consequence of this, you will often pass through many spells when the interchange of ideas between the two of you is beset with obstacles. This is because your intellectual rhythm will tend to be in an opposite phase to that of your partner's equivalent rhythm. When your rhythm is high, it will be much easier to express your thoughts and feelings clearly; your partner, on the other hand, might not be too receptive to your ideas and active state of mind. In view of these difficulties, try to build up a greater understanding of each other when the opportunity presents itself; communication is an important foundation in any relationship. In this way, you will be able to sail through those periods of misunderstanding much more easily.

INTELLECTUAL COMPATIBILITY 26-50%

Your compatibility is fair. On balance, your partner's intellectual rhythm will be in an opposite phase to your own mental rhythm. This thirty-three day cycle controls your thought patterns, reasoning powers, and the ability to assess quickly situations that instantly demand your attention. On the whole, you might well experience days when you just don't seem mentally to coordinate with each other. For instance, you might be enjoying an active state of mind, expressing lots of great ideas for the coming day; your partner on the other hand (being in an intellectually recharging phase) will find it hard to receive all those bright and new ideas. However, mentally aligned periods will occur. During these less frequent spells, take the opportunity to communicate your ideas and thoughts to each other more effectively, thus establishing a stronger relationship.

INTELLECTUAL COMPATIBILITY 51-75%

Your compatibility is good. This biorhythm influences your memory power, capacity for deductive reasoning, and the ability to relate your thoughts and intentions clearly to those in your company. On the whole, your cycle will tend to be found in a similar biorhythm position to your partner's. Hence, communication problems should not prove too demanding. There will be less frequent occasions when your intellectual biorhythms will be opposed. During these periods, your mental energies will not be matched, with the result that you might find stimulating conversations hard to come by. Use the more frequent spells of compatibility to increase your comprehension of each other's personalities, so that when any momentary disharmonies occur, they can be offset by this additional mental perception.

INTELLECTUAL COMPATIBILITY 76-94%

Your compatibility is very good. For most of the time, your intellectual rhythm will be in a similar phase to your partner's equivalent rhythm. This cycle controls your aptitude for mentally demanding tasks, logical reasoning powers, and synthetic thought processes. You will tend to pass through highs and lows at the same time, so you should have few problems when expressing thoughts and ideas to each other. When you are feeling mentally alert and capable of tackling something a bit more demanding, your partner will probably be in a similar state of mind. Therefore, shared intellectual pursuits are favourably aspected. Your low periods will create a more passive state of mind when less demanding mental activities should be entertained by both of you.

INTELLECTUAL COMPATIBILITY 95-100%

Your intellectual rhythm is exactly in alignment with your companion's intellectual rhythm. When your cycle is at a peak, your friend's will also be showing the same biorhythm condition. Also, you will encounter lows at the same time, as well as critical days. This rhythm influences your judgement, memory capacitiy, and general mental alertness. As your rhythms will always be in phase, the high periods will enable both of you to communicate with ease. The low periods will find each of you in a more passive condition that will need careful planning as rigorous mental work should be avoided during these spells. Critical days occur every 16.5 days, so seek outside advice on important matters during these phase changes if this is at all possible.

5 NATURE'S INNER NATURE

Just as body and mind are a unity, our inner time structure is part of our planet. What we may call a clock in ourselves may be a collection of oscillators, that tend to respond to the rhythms of earth. The time structure of our bodies is, after all, only partly within our skins, for we are open systems, unable to detach ourselves from the beats of this nature of which we are part.

BODY TIME – Gay Gaer Luce

This nature in which we live, this spinning orb in space, contains a myriad number of life forms. To the eye they all seem totally unconnected and separate. Yet, we find that both plant and animal kingdoms exhibit a sense of timing and a degree of sensitivity to their environment that tells us this can no longer be considered true. They constantly interact with each other and their surroundings in a way that reveals a far vaster biological and cosmological mechanism in action. Here you will discover other rhythms and sensitivities, not only in ourselves but also in various animals and plants, that reveal a perception that is hidden to all but the curious.

NUMBER CONSCIOUS SURF RIDERS

Look out, here they come! This variety of surf rider doesn't use a surf board, but rides the waves right up to the shore soon after the highest tides. If you are curious about this unusual animal then the coast of California will satisfy your inquisitiveness.

The animal is a fish called the grunion *(Leuresthes tenuis)*, well known to the inhabitants of this American state. The highest tides of the month occur at full moon and new moon. It is a few days after these lunar phases that these fish decide to ride the waves in order to survive. In the oceans of the world there are already too many predators that enjoy a feast on fish eggs, so this clever, survival conscious, moon-timed grunion has evolved a method of ensuring that her eggs do not satisfy all those hungry mouths beneath the waves!

A few nights after these tides, the sea is just below its highest mark. These egg laden fish swim up to the shore and deposit their eggs in the sand just at the water's edge. The timing for this spectacular event is critical: the eggs need about fourteen days to incubate, so if they are buried beneath the sands any later, they would be premature when the next high tide releases them from the safety of the beach. It is only through this remarkable sense of time that the grunion has been able to evolve this particular means of improving her survival odds. At any other point in the month the tidal position would be adverse, and all the young released on the next high tide would be untimely ripped from their sandy wombs.

From an animal that is able to reproduce in time with the moon, we now turn to other sea creatures that reveal sensitivity on a cosmic scale.

SOLAR ECLIPSE CHANGES CRABS AND FISH

In India, biologists decided to investigate physiological changes in two species of crabs, *Barytelphusa guerini* and *Oziotelphusa senex senex,* and one species of fish, *Channa punctatus* during partial and total solar eclipses from 15 – 17 February, 1980. In order to observe changes in the crabs and fish for both conditions, laboratories were employed at Palem (total eclipse) and Tirupati (partial eclipse).

Observations for physiological changes were made over the period of three days at eight a.m., three forty p.m. and midnight. (The middle day corresponded to the solar eclipse event.) The number of ions in a given volume of crab haemolymph and fish blood were estimated by using a technique known as flame emission photometry.

In the instance of calcium ions, the results were quite significant. At both stations all three species of animal portrayed a significant drop in the levels of these ions in their sampled fluids. This reaction suggested to the researchers that the calcium ions might well be moving out of the circulating body fluids and into the tissues of these animals.

For *Channa punctatus,* observations were only performed at Tirupati (the partial eclipse location). As already indicated, the calcium levels in the fish blood declined, but the sodium ions also followed the same pattern. However, when potassium ions were investigated, it was found that they started to rise and form a peak during the eclipse day – 16 February.

When both species of crab were studied, the following significant changes were observed – apart from the drop in calcium ions. At Tirupati and Palem, the levels of potassium were found to increase; analysis of the sodium ions revealed that they showed a marked decrease, but suprisingly, this only occurred at the partial eclipse laboratory.

Through the results of this experiment it can be seen that even these simple animals, that most of us credit with little sophistication or sensitivity, do in fact respond quite noticeably to changes in their environment due to solar eclipse.

Apart from the relatively rare occurrence of eclipses, the sun does not exhibit frequent changes in sunspot activity and flares. These alterations in the sun's normal stellar processes have been shown to cause changes in human behaviour as well. It is hardly suprising in view of the ionic fluctuations in *Channa punctatus* and the two species of crab that we also come under the influence of our main provider of light and heat. Calcium ions play an important part in the functioning of our central nervous system, and thus our behavioural responses. But, more of calcium ions later, and their effect on one ethnic group in particular.

SOLAR ACTIVITY CAUSES AGGRESSIVE BEHAVIOUR

In Douglas hospital, Montreal, Dr. Heinz Lehmann observed aggressive behaviour in some of his psychiatric patients. In an attempt to come to a greater understanding of these intermittent, hostile outbursts, detailed records were kept of any incidents that occurred. When several months of data had been accumulated, the incident days were compared with various factors that might give a clue to this sporadic, atypical behaviour. Changes in the weather, barometric pressure, staff shifts, dietary content, and visitor frequency all showed no correlation with these outbursts. It was only when these incident days were compared with information from the US Space Disturbance Centre in Boulder, Colorado that a definite link was established between sunspot activity/solar flares and undue patient excitement.

From the last two cases it can be seen that the sun has an influence on living systems on this planet in very real ways. It is no longer possible to think of our own lives, or those of our friends, as being totally separate entities that are unaffected by our environment. Through scientific research we are now beginning to understand the full meaning of environmental influences; these include changes from as far away as our moon, and still further to the sun – a distance of ninety-three million miles.

One influence that all of us expect is the daily rhythm of light and dark. (That is unless we live close to either of the poles where for months at a time we have to adapt to conditions of total light or darkness.) It is only in recent years that investigations have shown that visible radiation from the sun plays an important part in synchronising some of our biological rhythms. In instances of light deprivation, it can often be the case that some of our internal rhythms become irregular . . .

WOMEN CONCEIVE WITH A NIGHT LIGHT

In Helsinki, Dr Sakari Timonen investigated a large number of women over a period of five years for the incidence of cysts and other cellular abnormalities. After amassing a large amount of data on these conditions, the results were compared with the amount of monthly sunlight received in Finland during the investigation period. It was discovered that as the total hours of light per month decreased, pre-tumourous cellular activity increased, and the conception rate decreased. As the pituitary gland affects the menstrual cycle through initiating changes in the endometrium – the uterine wall – it was suggested that light levels might have an effect on the functioning of this master gland. This theory was furthered by research from the Rock Reproductive Clinic in Boston, where some infertility problems were successfully treated in a most unusual manner.

The interplay of light and the menstrual cycle gave rise to the idea that it might be possible to synchronize ovulation rhythms in infertile women who experience irregular periods. Dr Rock and Dr Dewan decided to study seventeen women who had infertility problems of this type. Their unusual synchronization regimen involved a lighting plan for these patients during the sleeping hours for just three nights in their menstrual cycle. The system was to ensure that these women slept from the fourteenth to the sixteenth nights of their cycle with a dim light on in their bedrooms. (The first day of their cycles was taken to be the day of menstruation.) Months later, when the menstrual records were investigated, it was considered a successful treatment: only two of the seventeen women had failed to exhibit the more normal twenty-nine day periodicity in their ovulation cycles.

Further tests were performed on another group of women but this time the nights of illumination were from the eleventh to the thirteenth days in their cycles. This altered timing in the lighting plan failed to have any effect on creating a more regular twenty-nine day menstrual cycle. It was, therefore, considered by Drs Rock and Dewan that the timing of these night lights was crucial to the reestablishment of a more fertile condition. As a result of this system, many of the women in the trial group became pregnant, much to the delight of all concerned. It illustrated that drugs and surgery were not necessarily the only methods.

Rhythm investigations in Wainwright, Alaska, into behavioural problems in Eskimos have also shown that light/dark cycles have the ability to alter human body chemistry considerably.

ESKIMOS FALL ILL IN THE DARK

People in polar latitudes often suffer from distinct changes in behaviour during the winter months, to the extent that they can be classed as definite illnesses. These dramatic changes in behaviour can appear quite suddenly, and may last from a few hours to a couple of months. The only constant factor in this type of malaise is that it always afflicts the Eskimos during the winter months.

Joseph Bohlen, who graduated at the University of Wisconsin, decided to explore this unusual condition in these polar-living peoples. Accompanied by his wife, he decided to study ten individuals, twenty-four hours a day – a mammoth task. At two hourly intervals the Bohlens visited each Eskimo's house and measured their hand grip strength, eye/hand coordination, blood pressure, pulse rate, oral temperature, and took a urine sample. Many scientific colleagues of the Bohlens wondered whether their researches would indicate that the Eskimos exhibited physiological rhythms that differed from people living in temperate regions. They considered that it might be possible for the body to adopt different cycles due to the long periods of continuous light followed by lengthy spells of unbroken darkness.

When the results of their 120 visits were analysed, it appeared that the Eskimos portrayed normal cycles in body temperature, as well as levels of urinary potassium. Blood pressure, eye/hand coordination, pulse rate, and hand grip strength all showed no significant deviations from that expected in temperate living humans. However, there was one rhythm that showed a distinct deviation from normal expected values. When the Eskimos urinated, they excreted up to *ten times* the normal amount of calcium than during the summer months. Gradually, as their frozen world changed from constant light to continuous darkness, their bodies gave up more and more calcium with the consequent upset of nervous function. It is known that anxiety neuroses are one result of too little free calcium being present in the body. Through the efforts of the Bohlens, this unusual circannual rhythm of calcium excretion was discovered which thus explained the mysterious winter afflictions of these northern, nomadic peoples.

PROFESSOR FRANK A. BROWN

Leaving human sensitivities to the environment aside for a while, there are many fascinating discoveries that have been made in the plant and animal kingdoms. One pioneer into biological sensitivity to the environment is Frank A. Brown, Morrison Professor of Biology at the Northwestern University. In the course of his researches he has revealed precise monthly rhythms that can be observed in many animal species. One case in point is the swarming of schools of Bermuda shrimp, their timing being synchronized with a particular lunar phase, or at least some other cycle that is of a lunar frequency. The advantages of the grunions' lunar/tidal sensitivity is apparent to all who appreciate the vulnerability of fish eggs. Likewise, the Bermuda shrimp has found the necessity of timing to ensure the survival of its species. This monthly rhythm might seem unusual and can perhaps be explained away by reference to other earth-bound zeitgebers, but there are also instances where animals swarm at a precise time of the year, phase of the moon, and time of the day. What gives them their sense of timing that is so important for the continuance of future generations? An animal that portrays this fine adjustment to time is the Atlantic fireworm. Through this massive gathering of a single species for reproduction, nature is providing the offspring with a greater chance of survival. The work of professor Brown has also touched on other very interesting areas.

Don't buy a barometer, buy a potato. . .

Surprisingly, he discovered that the common potato is a much more sensitive plant than one would initially assume. His investigations were directed towards the measurement of oxygen which all plants consume for the production of energy; a process called respiration. Within a hermetically sealed container he placed a plug of potato which contained an eye at one end. The plug was of a sufficient length to ensure enough nutrient for the developing shoots that were generated from the eye. Through analysing the gas surrounding the shooting plug, he was able to detect the amount of oxygen that the plug was using for its own energy requirements. The experiment had to be carried out in total darkness as any photosynthesis would produce oxygen and invalidate the results. The amounts of the gas that were taken up from the surrounding atmosphere were extremely minute, and so his measurements had to be very sensitive to appreciate any changes at all. Despite many teething problems that often trouble minute measurements of biological material, he was able to detect the expected daily cycle of respiration within the plug. But, in addition to this, he discovered that the potato altered its rate of respiration according to the daily fluctuations in barometric pressure. More remarkable was his discovery that the potato was able to predict the barometric pressure two days in advance by its afternoon peak in oxygen consumption, despite it being in an airtight container!

Other original work by Brown covered the effects of magnetism on animals, and for this he decided to study a marine mollusc called *Nassarius*. His experiment was simple, effective, and exhaustive, for he took a total of 33,000 observations of this slow-moving, slug-like creature that inhabits puddles along the sea shore. Each mollusc was placed within a large sea shell with some water in the bottom to prevent dehydration, and then it was observed for its exit orientation by means of an angled guide around the perimeter of the shell's mouth. This angle was recorded for each *Nassarius* and the experiment was conducted throughout the day with the time also duly recorded. When all of the information was collated, Brown and his research team found that the marine slug altered its orientation according to the time of day. In the morning they turned out of their shells going right. In the afternoon they tended to leave by left and forward directions. Looking at these rhythms of departure still further, Brown discovered that *Nassarius* was affected by the phases of the moon and also by subtle magnetic fields of the earth. Even more remarkable than pressure sensing potatoes and slugs following the points of the compass are some revelations made by the 'plant man', Marcel J. Vogel, of San José, California.

PLANTS THAT TALK, PLANTS THAT SCREAM. . .

Although Vogel has written about energy conversion in cystalline solids and is a specialist in solid state physics, one of his fascinating avenues of research involves the perception of plants. His work has taken him into areas that most of us would consider bizarre and eccentric, but the fact is that his curious, inquisitive mind has unearthed some interesting new viewpoints on how vegetable matter perceives, and responds to, its surrounding environment. Further more, not only has he realized a degree of sensitivity that was hitherto considered unlikely, but also that like animals that are sufficiently developed, some plants do exhibit behaviour that is very difficult to interpret without acknowledging that they possess a degree of memory.

I would like to mention three experiments that are really spectacular. The first examines a plant's perception of human emotions. The second demonstrates their environmental sensitivity to an even greater degree. The third experiment illustrates that plants have a memory.

In this first experiment, Vogel asked a sceptic to help him with an exploration into the ability of one of his indoor houseplants to pick up human responses. The invitation was simple – and almost embarrassing: just go ahead and talk to them! By way of introduction to this attempt of communication with the leafy world, Vogel eased his subject into the idea by explaining his theories on how the plant kingdom perceives electromagnetic variations in their surroundings, and through this, how that they are able to respond. This he intended to demonstrate to his bemused 'helper'; and this he succeeded in doing. Throughout this communication between man and plant, the plant leaves were wired up with electrodes that fed signals to a polygraph recorder. In this way, minute changes in the plant could be detected and instantly recorded onto a chart, and thus be compared with conversation that had just taken place in the room.

At points in their conversation, Vogel restated his beliefs that plants can tell what mood a person is in, and that they will actually respond to human emotions. (This is not a new idea to many 'greenfingers'. There are many horiticulturalists who talk to their plants, and even state that growth is improved when certain types of relaxing music are played in greenhouses over a home stereo system.) After the discussion Vogel and his helper went to the chart, and each time he expressed real doubts about the validity of Vogel's theories, the polygraph recorder described a noticeable spike and deviation from the rest of the trace. In many cases these doubts about the plant's sensitivity were not even voiced, just thought. Even so, the houseplant still detected his strong feelings of disbelief. I wonder what the recorder charted at the end of the session when all was revealed? The plant must have had a 'field day', with incredulous expressions and utter astonishment creating waves of emotional havoc. I bet those polygraph wires were buzzing!

Another experiment involving plant sensitivity seems even more notable, for it demonstrates that plants can sense changes in their environment, but at some considerable distance. In this instance, three plants were wired up to polygraph recorders, but they were each in separate rooms. At another end of the research building, an experiment of a much more painful nature was being conducted: live shrimps were being dropped into boiling water at random intervals, by a machine specifically designed for that purpose. The whole experiment was very accurately timed, in order that shrimp deaths could be plotted on the plant's response trace in each of the three rooms. Control experiments were also pursued so that the results of the research could be fairly interpreted as the plant's response to the timing of shrimp death. The results were always the same: when a shrimp died in the boiling water, each of the three plants gave a significant reaction that appeared on the recorder trace. As the timing of these events was random, this negated any possible cyclical variations of vegetable origin. This unusual sensitivity to other life forms has been attributed to plants being able to detect electromagnetic waves that emanate from all living tissue, and at the moment of death, the change of state of the organism produces a shock wave through the ether.

The last experiment illustrates that in addition to detecting destruction of life forces within living forms, plants also have the ability to remember events. This might even suggest some learning faculty if sufficiently sophisticated experimental designs were set up. This is where the 'murderer' enters our experimental stage.

Clive Backster set up an experiment in which he placed two philodendron plants in a room, side by side. He then chose six volunteers which had to enter the room in random order – and separately. However, one of the plants' visitors was instructed to destroy totally one of the philodendrons during his visit; the only witness to this act was the remaining plant. When the 'murderer' entered the room, the fated, unfortunate specimen was torn from its pot and then stamped into the floor until nothing remained but a tangled, crushed mass of roots, stems, and leaves. After all six people had been into the room, Backster wired-up the remaining philodendron in order to monitor its response to the visiting six. The moment of truth arrived when he asked them to file through the room, one at a time, as an identity parade. The solitary plant emitted a single, electrical scream when one of the volunteers drew uncomfortably close to its pot. *It had remembered!* It had detected the murderer's return to the scene of the crime.

From sensitive plants that are able to perceive and remember their surroundings to a degree that seems uncanny, we now turn to an animal that is invaluable to many of them. Without its ceaseless activity many plants would not be able to grow due to a lack of aeration and drainage in the soil. The animal is, of course, the humble earthworm.

EARTHWORMS GO QUICK, QUICK, SLOW

The work of Professor Brown on the magnetic sensitivities of *Nassarius* has stimulated similar types of experiment. It is one of these that I would like to mention now. Although the results are by no means clear-cut, they do indicate that other animals have the ability to sense terrestrial magnetic fields and alter their behaviour patterns accordingly.

The basic idea behind an experiment by M. F. Bennett was to record the length of time it took an earthworm to withdraw from a circle of white light; their speed of withdrawal being considered an indicator of their biological rhythmic state. The type of earthworm that was used for this investigation was *Lumbricus terrestrus,* and they were tested for reaction rates at midday (12 00 to 13 00 hours) and again in the early evening (19 00 to 20 00 hours). Specifically, the worms were placed within an enclosure that was diffusely illuminated by a red light at a subdued intensity of 0.5 lux. A white circle of light 25mm across was shined onto their head ends whilst they were positioned on a horizontal wooden platform with a means of removing themselves from this brighter (5 lux) circle. The whole apparatus was then placed within a further device that had the ability to generate magnetic fields twice that of the earth and also to reduce them to zero. The annelids that were subjected to these zero and twofold geomagnetic forces were compared with a control group that was kept in total darkness, constant humidity and temperature, and in the earth's normal magnetic field. A total of 5,000 single trials were performed and the results revealed several interesting trends, some of which were connected with the changes in magnetic field strength.

The experiment was conducted in two series, the first being during the autumn/winter when the control group was compared with experimentals who experienced a zero magnetic field. The second series was performed during the spring when the control group was compared with experimentals in a two-fold field. It was already known from extensive tests during the previous year that *Lumbricus* tended to move a lot faster during the evening observations by as much as twenty-five per cent, and this trend was reinforced by these two series involving the additional variable of geomagnetism, although only in the controls. During the autumn/winter tests, earthworms that were kept in normal magnetic conditions moved, on average, 19.1 per cent faster in the evening. However, when the laboratory annelids were observed, they were considerably affected by the zero magnetic field: they were markedly 'sluggish', showing only a 1.1 per cent increase in speed during the evening. For the spring series which compared normal magnetic conditions in the controls with two-fold fields in the experimentals, the differences were even more pronounced. The control group of earthworms raced along 36.3 per cent faster during the evening tests than during the midday, but the experimentals were hard pushed to muster more than an 8.1 per cent spurt over their lunch-time heats.

Although, as already mentioned, these results do not lend themselves to any resounding statements about exactly *how* these invertebrates react to magnetic fields that differ from the normal, they do reveal that they can sense magnetic changes which, in turn, modify their reaction rates to bright light. Further experiments with this earthworm where the anterior nervous system has been surgically altered have yielded similar reductions in their withdrawal speeds. This does tend to suggest that magnetic variations have a tangible influence on their nervous functions which then regulate their sensitivity and/or responses to the environment.

It is experiments like this with the lowly earthworm that reveal the foresightedness of Gay Gaer Luce when considering the range of sensitivities that are within the scope of human experience. Even simple invertebrates are open systems, unable to detach themselves from the beats of nature of which they are part, so how much more we –

as highly evolved, intricate examples of biological life – will be affected by the many daily and monthly changes that are continually being effected in our living world.

Are we, too, able to sense magnetism? Perhaps this smacks of fringe activities such as water divining and 'mystical' practices that all (excepting the misguided or foolish) consider as an imagined human ability since there is no scientific foundation. Or is there?

HUMANS AND MAGNETISM

There are many documented cases where dowsers have been consulted in order to assist in surveying land for the presence of water or even oil. Some appear to be more sensitive than others, but the fact remains that it is possible for 'gifted' individuals to detect the underground composition of the earth with a more than chance occurrence of success. Different methods are used, some employing a forked hazel twig held at arm's length, others require a pendulum containing the substance to be detected. Where the forked arrangement is employed, the twig will often dip violently when the operator is over an underground stream or significantly large area of water. With the pendulum, a change will be apparent in the oscillation, which is merely an amplification of the signals that the diviner has already received. Each individual has his or her own preferred methods that often seem intriguing and almost ridiculous to those viewing the procedure for the first time. So, is all this hocus pocus or are we more sensitive than at first thought? Can we measure any parameter that will reveal the communication of these changes?

Yves Rochard, a Sorbonne physics professor, investigated this controversial area of human perception. He discovered that if the arm of an individual is kept under tension whilst holding a long stick (as in dowsing), it is possible to detect accurately minute changes in the earth's magnetic field strength. Whilst his pupils set to work, rod in hand, exploring various terrains for the presence of water, he simultaneously took magnetic measurements of their immediate vicinity using a magnetometer that could detect extremely small alterations of magnetic flux. He found that subterranean water would be signalled by changes in field strength in the order of 0.3 to 0.5 milligauss. When these minutiae occurred, his pupils often detected water through their trembling hands. He also found out that this ability need not necessarily be a gift: through practice it can be acquired. His investigations went still further as he managed to conceal ingeniously electric coils within the ground that would upset the natural earth fields and hopefully be perceived by his band of marching, magnetic sensitives. Even average people were able to appreciate these small artificial changes that he had introduced, and so this successfully indicates that the human body is considerably more receptive to magnetic changes and might, therefore, be influenced in subtle ways through the immediate environment. Some of these subtle changes require painstaking research, and this is where the Max-Planck Laboratory has contributed much to our understanding of human responses to changes in geomagnetic flux.

The nature of this research was to investigate the sleep/wake cycle in human subjects within the confines of elaborate, experimental conditions. At this establishment, living quarters were constructed that were specifically designed for sustaining people for prolonged periods without any 'live' reference to the outside world. All time cues were strictly forbidden: no television, radio, or wrist-watches were permitted entrance into this sophisticated, living complex. In addition to this, the only light source was artificial, which was activated by each volunteer and was not time-switched to any routine. In return for their relinquishment of all earthly time cues, each volunteer was provided with comfortable living and sleeping quarters, together with any personal effects that would make them feel more at home in their remote and estranged abodes. In one experiment, eighty-two people were assessed for their sleep/wake patterns, but unknown to them, although their rooms looked identical on the inside, they were different in one vital aspect. Some of the time-frozen chambers were heavily shielded against any terrestrial magnetic and electrical fields by generous metal plating on the outer surface. In effect, these invisible influences were blocked out. Of these eighty-two 'guinea pigs', thirty-three were studied within chambers that were bare to these earth-borne pulses. A very interesting pattern slowly began to emerge, for it took several days for volunteers to settle down to a stable routine without the usual discipline of earth time cues, and even longer for the researchers to obtain an average figure of their sleep/wake cycle periods.

Drs Aschoff, Wever, and Poppel discovered that the subjects who were passing 'time' in the unshielded compartments had an average day length of 24.84 hours. But the results from those cocooned within metallic envelopes revealed that they had adjusted to a longer day of 25.26 hours – about twenty-five minutes extra. There was also one other feature in this series of tests: the subjects in the exposed chambers exhibited less internal desynchronization. This meant that certain parameters such as peak urine production tended to occur during the expected period just after a prolonged rest or sleep. Contrastingly, the volunteers in the shielded rooms portrayed kidney excretion rhythms that were out of phase with their physical activity cycles.

It can therefore be concluded from this experiment that magnetic and electrical earth fields not only affect the total cycle time for sleep/wake patterns within these subjects, but also alter other body rhythms. These results do tend to suggest that we are dependent, to some extent, on geomagnetic influences to keep our bodies in time. It has been discovered that in some cases where individuals have a previously inexplicable illness, their internal biological rhythms were desynchronized. They were 'out of tune', much like a car engine whose timing is no longer in step.

Within our complex bodies, there is an appointed season for each activity, and when we become out of phase within ourselves, it is then that our system becomes stressed, with a corresponding display of malaise. For instance, our body temperature follows a distinct rhythm, fluctuating between ninety-six and ninety-nine degrees in a healthy person. During the early hours of the morning it will be at its lowest, proceeded by a gradual rise until it peaks later on in the day. Assuming that a person tends to drink equal and regular amounts of liquid when awake, they will also exhibit a rhythm in the volume of urine that is excreted: there being more produced in the morning than in the late afternoon and early evening. We also have rhythms in the excretion of sodium and potassium ions that are found in urine, and as these are very easy to analyse and quantify, they become rapid indicators of our cyclic state.

In 1965, one of Dr Aschoff's volunteers exhibited this out of tune condition, for during his prolonged spell of isolation, his sleep/wake cycle had settled down to 32.6 hours. Although this might initially seem to be perfectly acceptable – after all, his time was his own, with nobody to dictate any form of routine – some of his other body rhythms were not following suit. This was where his problems began: the body temperature and urine excretion rhythms both stabilized at 24.7 hours. These two differing frequencies then dictated periods of harmony between the three rhythms, and also spells of total discord.

It is very interesting to note that this subject's diary always indicated that he was feeling alert and much more healthy when all three of these rhythms were momentarily in phase with each other. This points to the fact that one health factor is a harmonious relationship between all of our biological rhythms. These can vary from cycles which complete themselves in minutes, to days or even months, and longer. When any of these rhythms are opposed to each other, we can expect to be feeling less than our usual self – jet lag is a classic example. These events must also be considered in conjunction with our three supradian rhythms, so now for another look at critical days, and how some people have reacted to them.

6 RHYTHM OF THE MOMENT

By the law of Periodical Repetition, everything which has happened once must happen again and again and again – and not capriciously, but at regular periods . . . the same Nature which delights in periodical repetition in the skies is the Nature which orders the affairs of the earth.

MARK TWAIN

We have seen how that biorhythm has the ability to affect our moods, physical vivacity, and intellectual concentration throughout our lives. We have a biological resonance within our skin and bones that is nature-born, necessary, and regular – like the rotation of the earth around the sun, the changing of the seasons, or the always expected succession of day after darkness. Dr Wilhelm Fleiss was the first explorer of these harmonious, life-giving forces, who postulated that their effect might be a significant factor in certain types of mortality where illness is existent and progressive, culminating in a breakdown of body functions that are necessary for the continuance of life. During these conditions, the individual becomes appreciably more sensitive to *changes* in these inner cycles, and so biorhythmically triggered trauma become all the more probable, and possible.

During this chapter I will concentrate on critical days, graphically illustrating the process that is in operation. The biocharts will specifically depict conditions that are existent at the time of death or other notable event, and the case histories that I have chosen are those involving illness, rather than sudden externalized trauma that prove fatal. The exception to this is, surprisingly, suicide. Linked with this I have also touched upon the criminal mind as both states can often be attributed to unstable thought patterns that are potentiated by critical events. There are some interesting stories to be told along the path of criticality, providing ample food for thought. I suspect that your conclusion will be as provoking as mine.

SUICIDE

Although one cannot sensibly include other types of death due to sudden accidents that have been inflicted upon the individual – such as motor vehicle fatalities involving passengers, travellers caught up in the maelstrom of an airplane crash, or disasters for those voyaging at sea – suicide must be included within the bounds of biorhythm influence. Most people would agree that when a person is contemplating suicide, the balance of their mind is temporarily disturbed; they might be extremely depressed or overwhelmed with fear. Within their psyche, it is often a very negative complexion that colours their life horizons and opportunities with an impenetrable gloom. For them, there is no point in living any further. Taking these self-inflicted mortalities, Kitchinosuke Tatai has shown a significant case for biorhythm influence. He lists several famous people, some of them Japanese, that took their lives close to, or coincident with, physically and/or emotionally critical days in their biorhythm profiles. These critical day combinations have been shown to have a greater effect on the human mind and body; especially the physical/emotional double critical event. The following instances of suicide make the point very clearly:

Novelist Osamu Dazai committed suicide only one day after passing through a double critical in the physical and emotional cycles, followed by an intellectual critical. Another famous novelist, Ernest Hemingway, shot himself on 2 July, 1961, on an emotional critical – and only one day away from phase changes in the other two rhythms. Henri de Monteran shot himself one day after a physical critical, additionally, both the emotional and intellectual rhythms were at their lowest points. Motion picture director Akira Kurosawa ended his life in December, 1971, whilst passing through an emotionally critical day. A physical critical had occurred only three days before, and if he had lived another day, that would have brought him into an intellectually critical condition! Another novelist, Ryunosuke Akutogawa, committed suicide when experiencing a physically critical day. Both intellectual and emotional rhythms were at an energy nadir. Finally, another novelist who *planned* a suicide during an *intellectually* critical day – and died by his own hands only four days later when physical and emotional cycles were at mini-criticals. This last case history illustrates how that reasoning powers are sufficiently diminished during intellectual phase changes, and thus the contemplation of one's own self-annihilation becomes that much more 'reasonable'. A sobering thought indeed.

All six suicides depict the artistic, creative personality, so perhaps this suggests that that temperament is more inclined to instability at a phase change. This could be purely coincidental, and I tend to think so. Unfortunately, Tatai does not interrelate a larger sample of biorhythmically analysed suicides with profession, to detect whether this might be true or not. A lot more work needs to be done in this important area of human behaviour. Why not start your own research into suicides and biorhythm? You have all the necessary tools at your disposal.

CRIMINAL BEHAVIOUR

From many other cases of suicide, it does appear that mental stability is influenced by biorhythmic phasing. This discordant state of mental affairs is one of the key contributing factors when considering deaths of this order. Similarly, another area that has come under the scrutiny of biorhythm analysis is that of criminal behaviour – again, often associated with abnormal mental states. Whether the weakness is of the body, or resident within the unseen mind, it is evident that phase changes do upset the human status quo, and in the wake of these changes one can observe behaviour that illustrates this inward process. One of the most spine-chilling bioprofiles that I have studied is that of a multiple murderer who terrorized Japan during the October and November months of 1968. All who harbour any doubts about biorhythm read on. . .

The notorious crime series began on 11 October, (one day after this man's emotionally critical day) when he shot a guard at a Tokyo hotel; the bullet wound was fatal. His desire for further bloodshed was soon satiated, for only four days later – when his body went through a physically critical day – he shot and killed a policeman at the Yasaka shrine in Kyoto. After another seven days, his emotional cycle passed from negative to positive; thankfully, no killing was reported . . . but time was fast running out for his next unsuspecting victim of only five days hence. So, on 26 October, the double murderer shot a third time, and his prey was a taxi driver. During this third killing, his mind and body were destabilized by a double critical in the physical and intellectual cycles. Finally, ten days after this on 5 November, he murdered another tax driver. On this occasion he was 'sandwiched' between physically and emotionally critical days that were only forty-eight hours apart. He was, therefore, within the phase switching, unstable zones of both cycles. The police were unable to identify and arrest this murderer for another four months, but their efforts were at last rewarded by another violent outburst on 7 April, when the man was critical in his emotional cycle. All incidents occurred when he was either physically critical, or critical in another biorhythm and physically high at the same time. For those with criminal tendencies, it has been found that an emotionally critical day coupled with a physical high is particularly dangerous, and merits increased supervision of the individual. This is due to a rapid discharge of body energy in an uncontrolled manner as the critical emotional rhythm has influenced sensitivity thresholds and immediate environment perception. With these lowered thresholds it is no surprise that an already defective and destructive personality will maximize all available energy, and bring it into malevolent operation.

BIOCHARTS

The following ten biocharts illustrate conditions that were in existence at the time of death for various famous people (with the exception of the Japanese murderer). The main focus of attention will be upon the influence of critical days, although other conditions must be taken into account as well, such as mini-critical days. Triangular arrows on these biocharts indicate the day of interest.

George Thommen noted that Clark Gable's first heart attack occured on 5 November 1960, when his physical rhythm was critical. He subsequently predicted extreme caution for the 16th, and as you know, Clark Gable died on this day as Thommen had foreseen.

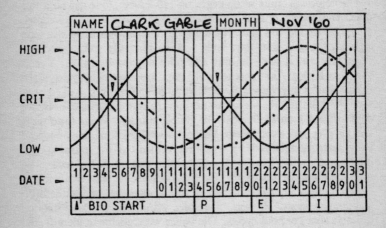

Just after midnight on 28 June, 1967, Jayne Mansfield had a fatal accident whilst driving along a New Orleans highway as her car switched lanes and crashed into an oncoming truck. There were news reports that she had been drinking. At the time she was physically critical and within twenty-four hours of emotional and intellectual critical rhythms as well. Although this is not compressed enough to be considered a triple critical day, all three rhythms were certainly having their influence.

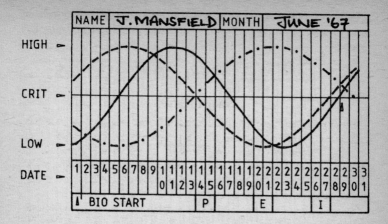

The infrequent triple low marked the death of Violet Carson on 26 December, 1983. She was well remembered as Aunt Vi on BBC's 'Children's Hour', and also as Ena Sharples on 'Coronation Street'. She fell ill with pernicious anaemia in 1980 and had an operation the following year from which she never fully recovered. In a debilitated condition, the triple low can have equally destabilizing effects as the more 'popular' critical days.

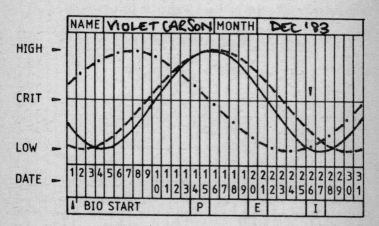

The Japanese murderer committed all of his crimes on critical days or semi-caution days. The first was on a semi-caution day in the intellectual rhythm. The second was on a physically critical day. The third was between phase changes in the physical and intellectual rhythms. The fourth (not on the biochart) occurred when he was experiencing semi-caution days in both the emotional and physical cycles.

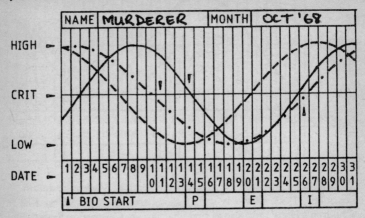

'Sandwiched' between emotional and intellectual criticals, Arthur Askey died of shock on 16 November 1982, after his second leg had been amputated only a few weeks before. He also had a history of heart trouble. He was one of the first comedians to exploit radio for comedy with the commencement of 'Band Wagon' in January 1938.

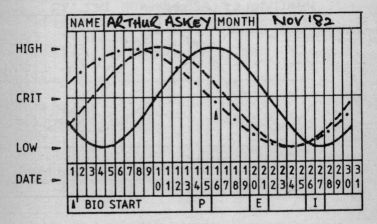

Elvis Presley had just recovered from a double critical day in his physical and intellectual cycles when a mini-critical event in the emotional cycle occured on the 16 August 1977. The same day he died.

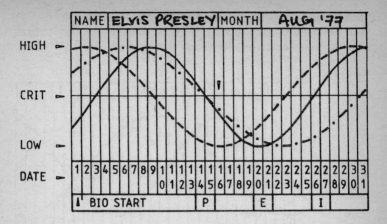

Leonard Rossiter collapsed and died of a heart attack during a stage performance of 'Loot' on 5 October, 1984. He was a well-known actor, especially for his portrayal of the seedy landlord Rigsby in 'Rising Damp' and also for the leading role in 'The Rise and Fall of Reginald Perrin' – both British TV comedy series. The chart shows that his body was well within the influence zones of emotionally and intellectually critical events at the time. (Almost a double critical day.)

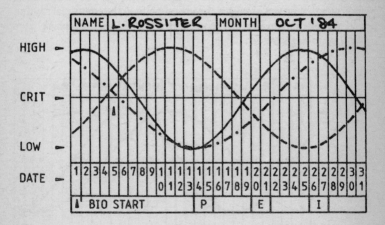

Bing Crosby died of a heart attack shortly after playing a round of his favourite game, golf. Looking at his chart, you will see that he was experiencing a semi-caution day in the physical and intellectual cycles.

The emotional high might have contributed to this as well, by making him too excitable. Any over exertion coupled with a physical critical event will only exacerbate an already strained body.

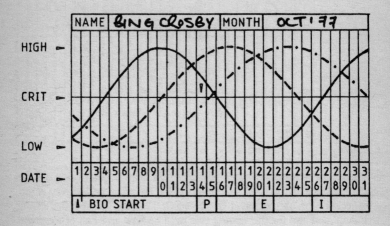

After a long struggle with cancer, Reginald Bosanquet died on 27 May, 1984. His jovial personality made him one of the most popular newsreaders on 'News at Ten', which started in 1967. His chart indicates that he was almost at the lowest point in the emotional rhythm and that one day before, his physical rhythm had changed phase from positive to negative.

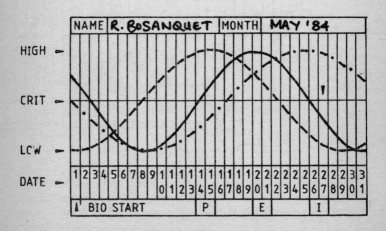

98

Edward, Duke of Windsor, died on the most unstable biorhythmic event possible: the triple critical day. For quite some time he had been suffering from cancer of the throat, and he passed away at his Paris home at 02.25 on the morning of 28 May, 1972.

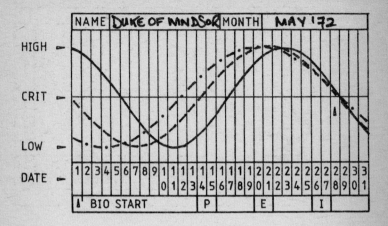

7 STRESS AND RELAXATION

Stress is largely what you make of it. What is stress to one person may be rest to another . . . So you see it is no good saying to people in general, 'You must avoid stress', because the man who thrives on stress would be driven mad by the monotony of a life without stress. At the same time there are others who are trying to escape from responsibility the whole time. Others again, who may lead quiet lives with no responsibility would turn everything into stress.

HOW NOT TO KILL YOURSELF – A Family Doctor

Stress is part of everyday life, and the antidote to much of it is to learn how to relax properly. Often, many of us are needlessly struggling under the burden of bodies worn down by it. How many of us have discovered that our bodies are actually suffering through harmful thoughts and negative attitudes? There was a time when 'mind over matter' theories were rife – and always laughed to scorn. But now we are beginning to look at the mind / body issue more seriously because medical research is proving that with our thinking we have the power to heal ourselves or become the breeding ground for endless self-inflicted maladies.

Firstly, let's explore the mind, and how it is capable of affecting us, together with a quiz that will unearth hidden attitudes within your personality and discover those innermost thoughts. All will be revealed. Secondly, your most vulnerable moments whilst under stressed conditions are, as you have already witnessed, critical days. I will deal with these, and an easy to follow, quick method for ensuring a more relaxed outlook on your day – even with momentary instability in your life cycles. (For those of you that are accident prone – and look upon the critical event with trepidation – remember those accident statistics from Japan: in almost every case, an awareness of critical events was enough to prime the body's corrective mechanisms against any unwelcome incident.)

MATTER UNDERMINED?

If you look on the black side, it could be more than your vision that changes. . .

Blood cholesterol levels can be raised as much by anxiety as by

101

indulging in high fat foodstuffs such as dairy produce and fried foods. In a five year study based at the Northwestern University Medical School in Chicago, hundreds of coronary-prone patients were monitored because their smoking, over eating, or high blood pressure had placed them into a high risk category. During this period, each of them was asked to record their degree of tension. Cholesterol levels were found to increase each time they suffered a nervous upset, irrespective of eating habits, body weight, or daily consumption of cigarettes. It seems, then, that stress reactions push up blood cholesterol, so what makes it come down? Believe it or not, a group of rabbits have the answer.

One experiment into the health of rabbits with a high cholesterol diet gave some interesting results. Two groups were set up, both had the same feed, but one was frequently taken out of their cages and given a lot of affection and handling by the researchers. The result? Those that were kept in isolation were *less* resistant to high cholesterol related illness. Affection and love had the ability to counteract what would normally be considered a very unhealthy diet. So the next time you're stressed – and your blood cholesterol is about to peak – tear down those mental barriers: your closest friends may have more of a healing influence than you had initially realized.

In a study of terminally ill cancer patients, it soon became apparent that their mental attitudes had the ability to alter considerably their life expectancy. Dr Achterberg has discovered that those who outlived their fellows were predominantly positive thinkers. They never gave up hope. They had a strong belief in their own abilities, and were unable to accept the fact that they were invalids. They almost denied the existence of their condition. From her in-depth work with these patients and the effect of the mind on the body, Dr Achterberg says, 'Every thought affects the body in some way, and every physical movement or change is accompanied by some kind of mental alteration.'

If you are the type of person who tends to 'bottle up' your emotions, then don't! Dr Steven Greer, a senior lecturer at the London Institute of Psychiatry, has found that women who are unable to express their anger are far more prone to breast cancer than those who let their feelings show. 'Getting something off your chest' by unburdening a problem to a close friend takes on a whole new meaning. Let's hope that we remember it.

After thirty years of research by American experts, it now appears that the body's cancer defence mechanisms *can* be broken down by the individual's inability to express his or her deepest creative impulses and emotional needs. At present, many scientists feel that our emotions are being transmitted to each cell of our bodies by hormones and other chemical messengers, and that habits of the mind and 'heart' are etched upon these smallest units of our mortality. Putting it another way, we are continually 'programming' ourselves for health or disease. Through wrongful, destructive attitudes we eventually undermine the very systems within ourselves that are specifically designed to keep us in the peak of health.

Feeling angry and denying expression to your creative impulses are not the only states that have been shown to induce chronic illness. A holistic medical centre in Phoenix, Arizona has revealed that many of its arthritic sufferers have long term emotional problems that are closely linked with this painful, immobilizing condition. Hatred, resentment and even frustration are the most frequent offenders. Through skilled counselling, many have learned the art of forgiveness that paves the way to effective, natural, healing processes. The Centre has observed dramatic improvements in patients' conditions by their new-found approach to life.

When you want to opt out of life's pressures and say goodbye to all that responsibility – but end up doing nothing about it – you may be giving yourself heart disease. Psychologist Kenneth Pelletier believes that many heart patients have this type of approach to their work or home environments. Eventually their life spans are cut short by subconscious transmission of these feelings that tell the heart it no longer serves any useful purpose; the body responds by shutting down its cardiovascular system.

THE POWER OF POSITIVE THINKING

In the West, awareness of mind/body interaction within the medical profession emerged in the 1920s with the conception of autogenic training by Dr Johannes Schultz. These training sessions are aimed at reducing the effects of stress reactions within the mind and body through relaxation exercises coupled with positive mental attitudes. They often start with the individual sitting or lying down in a comfortable position. This is followed by making conscious suggestions to alter the functioning of various parts of the body. For instance: my heart is beating slower; each limb is becoming heavier, tension free, and warmer; my breathing is getting deeper and much more relaxed; the muscles in my neck and shoulders are releasing all of today's tensions. In the mid-1980s many of these principles are expounded on relaxation tapes and self-hypnotic courses because their efficacy has been proved time and time again. In many alternative health centres in England, Schultz's methods are still being used because they work. As strange as it may seem, it *is* possible to just think yourself fit. . .

Research programs have determined that individuals who have been engaged in autogenic training for six months or more reap numerous physiological benefits; they become more healthy. Zen meditation (which is often looked upon as a bit cranky) is not excluded here, for it has also been found to have similar influences. All these changes signal that the body is better equipped to handle stressful situations. To name just a few, training normalises the amount of adrenocortical hormones that are released when under stress, reduces the amount of cholesterol in blood serum (the rabbits' handlers had a similar effect), and, more noticeably, lowers breathing and heart rate. Most important of all, it helps to create a calm and peaceful feeling in the trainee. This is a vital part of the process because long-term stress related illness is often due to a state of over-arousal. This calming, pacifying technique is, therefore, most effective in neutralizing these cell-trapped tendencies.

We all know that anxiety causes tension of some sort, and it is usually the body that gets the full brunt of it. It might reveal itself as muscle tension, trigger off a blinding migraine headache, or have you reaching for a packet of antacid tablets. Biofeedback, in particular, has demonstrated in an exciting way that the human mind is a veritable powerhouse of healing energy – when channelled precisely and correctly. The principle behind biofeedback is that the individual is conscious of specific body functions with the aid of instrumentation. Through changes in mental attitude he or she is able to bring about the desired physical response. It takes autogenic training one step further because the individual is constantly aware of what is happening beneath his skin.

A biofeedback trainee might be wired up in such a way that they instantly know their heart rate by looking at a bank of flashing lights or by listening to changes in sound signal from a loudspeaker. Muscle tensions are also relayed through sound: when a particular muscle is tense, the pitch is high. When it is totally relaxed, the pitch is at its lowest. It is not hard to imagine the success of this particular technique when eliminating wrongful postures or bad muscle habits that have developed through long-term stress. Other feedback systems are skin resistance and temperature; both are sensitive indicators of unease that quickly yield promising results. With a temperature probe attached to the forehead and another to the finger, a mind/body trainee will be told to cool his forehead temperature in relation to his finger temperature. This particular technique soon brings a relaxed response, and it is interesting to note that early autogenic training sessions were often completed by visualizing this forehead-cooling process.

To quote Gay Gaer Luce and Eric Peper: '. . . biofeedback promises to return us to a more holistic kind of medicine in which the patient will acquire more responsibility for, and power over, his own health, no longer finding himself treated as a defective organ, but as a person in a context, with a life style and habits that affect his own body. . .'

Even, more exotically, one's brain waves can be broadcasted: the alpha and theta waves being linked with deep relaxation states. Those that are able to switch into these waves through listening to a sound feedback of their brain rhythms soon become masters of their own consciousness. These states are often referred to as a half-sleep where the individual is in a tranquil frame of mind, and sometimes more creatively receptive as well. Biofeedback is just one system that is able to confirm the powers of the mind, for good or for ill, as the following case history reveals.

Dr Pelletier gave an example of a woman patient of his who suffered from a painful condition known as spasmodic torticulosis, where the head and neck were twisted over to the right shoulder. As a result of the complaint, she was unable to walk and her treatments had included the following: cortisone injections, chiropractic manipulation, traction of the neck in hospital and wearing a surgical collar. None of these gave any relief to the problem. But when she was wired up to a biofeedback machine that monitors muscle tension (sometimes known as an electromyograph) it registered an intense level of muscular activity. By careful questioning, Dr Pelletier discovered the root cause of this condition: each time the woman had a painful spasm, she felt full of guilt and shame. Further probings revealed that her symptoms started about five years before – she had commenced an affair with a younger man at that time. As soon as she faced up to her emotions and expressed them, the biofeedback monitor registered normal muscular activity. Her mind released the secret; her body released the tortured muscles. The healing process had begun.

Now, more than ever, the medical profession is realizing the full extent of psychosomatic disorders. We can play as much a part in the healing process as the healing practitioner. The recent formation of the British Holistic Medical Association heralds the way to more effective thinking about these processes. Alternative medicine increasingly points out that the quickest and safest route to rejuvenation is to allow the body to heal itself, rather than to resort to powerful drugs that are foreign to our natural biochemistry.

The tragedy is that we have been seduced into looking outside ourselves for the healing balm: within our own minds we have the most powerful antidote to ill health – and that includes the host of stress-related symptoms. That the mind has a tangible and immediate effect on our bodies is not a new or novel idea, but its enlightened application to the realm of self-healing is one that cannot be bypassed any longer.

This is an appropriate point to test yourself to see how positive or negative a person you really are. Biofeedback proves that this is very important to general health; cultivating helpful and constructive attitudes about your self-image, job situation, or home life will greatly aid the healing process. Here, then, are some questions that will probe your personality and reveal those hidden attitudes. Score each question as follows: 1 = Never, 2 = Rarely, 3 = Sometimes, 4 = Often, 5 = Always.

1 Do you picture your life style in the coming year as being more comfortable and content than at present?

1   2   3   4   5

2 Do you feel at ease with friends' jokes that have a little 'dig' at your idiosyncrasies?

1   2   3   4   5

3 When you have to face a change of plans unexpectedly, are you swift in looking for hidden advantages?

1   2   3   4   5

4 If an *equal* at work is better than you at your job, would you endeavour to learn from them and improve?

1   2   3   4   5

5 If a doctor told you that you had cancer, would you try to fight it 'tooth and nail', reprogramming your mind and body for health, believing in a total remission?

1   2   3   4   5

6 When you reminisce over the past year, do you dwell more on the high points than the disappointments?

1   2   3   4   5

7 Do you find it easy to compliment your spouse or lover?

1   2   3   4   5

8 If you think that some friends are talking about you, do you assume that they are being complimentary?

1   2   3   4   5

9 If a cashier short-changes you at the supermarket, do you have the confidence to go back and argue your case?

1	2	3	4	5

10   Do you take it as a compliment if a close aquaintance starts to mimic your life style?

1	2	3	4	5

11   If you trip up in front of friends and go 'flying', are you able to laugh it off as you see the funny side of it?

1	2	3	4	5

12   If someone finds fault with your work, can you differentiate between a constructive remark and jealousy or sarcasm?

1	2	3	4	5

13   Do you look forward to being in the company of people that are noted for their success in life?

1	2	3	4	5

14   Do you visualize yourself as a happy and relaxed person?

1	2	3	4	5

15   Are you an admirer of beautiful people and beautiful things?

1	2	3	4	5

16   When you have had an argument that has proved you're at fault, do you quickly apologize?

1	2	3	4	5

17   If you look up and find that a stranger is exchanging glances with you, do you think that they are attracted to you?

1	2	3	4	5

18   If someone breaks an appointment with you are you surprised?

1	2	3	4	5

RESULTS

78 +   Remarkable – you are a very positive thinker indeed!

72-78   Excellent – your positive thoughts are undeterred by most situations.

66-72   Good – on the whole, your outlook is predominantly positive.

60-66 Fair – negative and positive attitudes tend to balance each other.

60 and Poor – you need an injection of positive vibes! Give your
below attitudes a shake up, and your mind a spring clean. Then see your life change for the better.

RELAXING AT THE CRITICAL MOMENT

With careful planning as a result of consulting your biochart for critical events, and your body charts for cyclical patterns of reduced performance, you will be able to reduce the effects of high risk periods in your day-to-day life to a bare minimum with the implementation of an effective relaxation schedule. This simple-to-follow schedule combines well established techniques in autogenic training and relaxation breathing, plus a few visualization methods. All of us suffer from general stress at some time or other, but by using these two systems you have the distinct advantage of being able to pinpoint, in advance, those days when you might react adversely to high pressure situations. With practice, these techniques will become a part of your normal routines for coping with tension, especially when critical. If you want to include them as part of each day – first thing in the morning and last thing at night – you will be making a very positive step towards reducing stress reactions within your mind and body.

Before commencing with the relaxation schedule it is important to study the science of correct breathing. The way we breathe has a dramatic effect on our thought processes and also our state of physical health. Dr Claude Lum, a chest consultant, has conducted in-depth research into hyper-ventilation which reveals that chronic over-breathing can lead to anxiety states. Conversely, he feels that learning how to breathe deeply can, *without any other system,* be highly effective in reducing stress related illness in very tense individuals. A high degree of tension and anxiety is also thought to link with certain types of heart condition. If you have any doubts about these findings then proceed with this method and discover the benefits for yourself.

There are many systems that can be adopted, but here I will describe the complete breath because so few people use their lungs to full capacity and to best advantage. At all times breathing should be performed through the nose, with the mouth closed. This is because, unlike the mouth, the nasal cavities are able to filter incoming air and warm it before it continues its flow into the lungs. Practise this sitting in a straight-backed chair as it will be combined with other techniques later on. It is desirable to remain mentally alert, and yet as physically relaxed and comfortable as possible throughout. Ensure that your spine is straight and also erect. If any clothing is restrictive to natural breathing movement (neckties and belts), then loosen them until you have the freedom required. If possible, try to conduct this technique in a quiet place where you will not be disturbed, with a good circulation of fresh air – but don't freeze to death!

THE COMPLETE BREATH

First of all, close your eyes – this will help you to concentrate on the breathing process. Take your wrist pulse with the fingers of your other hand. This is to time your breathing pattern. What you should aim at is four heart beats for an inhalation, two heart beats for a rest, and four heart beats for an exhalation. If you find this too fast, it need not be strictly adhered to. You might prefer 6, 3, 6 or 8, 4, 8. However, keep to the ratio indicated. With practice you will be more attuned to your heart beat – especially in quiet surroundings – in which case you can dispense with taking your wrist pulse, thus freeing your arms which you can then place by your sides. The complete breath is from steps (1) through to (7). Steps (1) to (3) take four heart beats, (4) takes two, and (5) to (7) take four.

1 *Lower Inhalation* – Relax your abdominal muscles and inhale by means of lowering your diaphragm only. This might take some practice at first, as the rib cage should not move. However, you will be quite surprised to discover the amount of air that can be drawn in at this stage.

2 *Middle Inhalation* – Smoothly follow on from step (1) by allowing your rib cage to move outwards but not upwards. This phase will come naturally to you as it is more often used.

3 *Upper Inhalation* – Smoothly follow on from step (2) by allowing your rib cage to continue its natural movement upwards and also outwards. Continue this phase until you are unable to draw any more air into the lungs, but do not strain.

4 *Rest* – Hold your breath for half the number of heart beats that you feel comfortable with for the inhalation and exhalation phases.

5 *Upper Exhalation* – Start the exhalation by allowing your rib cage to drop. Do not contract your abdominal muscles or *force* any air out of your lungs.

6   *Middle Exhalation* – Smoothly follow on from step (5), letting your rib cage move inwards. Again, do not try to contract your abdominal muscles.

7   *Lower Exhalation* – This takes place when you are unable to expel any more air from your lungs by movement of the rib cage alone. Contract your abdominal muscles and allow the diaphragm to rise up into the thorax, thus expelling any residual air left in the lungs.

This breath is very simple and fully exercises your lungs. It is also a cleansing breath due to the hold in step (4). Try about twenty complete breaths of this kind to get into the natural motion that is required for further relaxation. In time you will proceed automatically by sensing your heart beat and following total inhalation and exhalation naturally.

It is important not to overdo this procedure by speeding it up, as it is possible to over-oxygenate the blood with the result that you will get light-headed, so keep to your heart beat rhythm and discover the peaceful and relaxing sensation of the *COMPLETE BREATH*. There are no hazards in this procedure provided that you follow the principles carefully. During the cycle, although each phase has been described, try to make a smooth transition from one to the other. Jerky movements will only perpetuate any tensions that you are already carrying. Make a point of monitoring your arm and leg muscles during the breath – many people tense up their limbs when going through the cycle. *MAKE SURE THAT YOU ARE RELAXED ALL OVER.*

When you feel confident about the complete breath, and are not putting all of your concentration into correct movement synchronized with heart beat, proceed with the following program. By breathing automatically, you will then release your mind for other tasks in this deeper relaxation technique.

The relaxation program is in three parts that are designed to relax you in degrees. The first stage involves muscle stretching which will ensure a more comfortable posture whilst sitting. This is vital, for if you are aware of any muscular tension, this will distract you from the deeper levels of the process. The second stage comprises complete breathing. This will stimulate and yet calm the mind, which will aid in the effective implementation of the third stage. At the same time, the rhythmical breathings will also relax muscles in your abdomen and thorax. Thirdly, having relaxed through stretching and breathing, conscious visualization and suggestion techniques are employed to reduce mental tension and build up more positive attitudes that are invaluable when encountering stressed situations.

RELAXATION PROGRAME

STAGE ONE *(time approx. 3 mins)*

During these four exercises sit in a straight-backed chair and breathe freely. Keep both feet on the ground, but slightly apart.

1 *Leg Stretching* – First of all, place your hands on each knee. Raise both feet up on tip-toe, your knees should also be raised by a few inches. Try to press both feet into the ground, but do not press onto your knees with your arms. Keep this position for a couple of seconds, then relax. Allow the soles of both feet to meet the ground. Repeat this procedure three times.

2 *Spinal Twist* – Let your hands fall down by your sides. Raise each arm until it is at right-angles to your body. (When viewed from in front or behind, your body and arms will form a 'T' shape.) With your arms held in this position, turn your whole body to the left so that your left hand swings behind your back. Allow your head to follow this movement. Reverse the movement and turn your body to the right so that your right hand swings behind your back. Allow your head to follow this movement. Repeat this 180 degree arm-swing three times.

3 *Back Stretch* – Raise both arms in front of you until they are horizontal. Interlock both palms and then turn them to face outwards away from your face. Raise both arms in this position until they are vertical. (You will probably want to take a deep breath here.) If you can, continue this swing until you are arched slightly backwards. Allow your head to go back as well, and keep your teeth together! You should feel a strong sensation of stretching in your neck and chest muscles. Hold this backward-stretching position for a few seconds, and then return to the start position. Repeat this back stretch three times.

4 *Neck Rotation* – Let each arm fall loosely by your side, and start this final exercise by allowing your head to come forward with your chin resting on your chest. With the head, describe a circle in an anti-clockwise direction that takes it over the left shoulder, backwards, over the right shoulder, and forwards. Repeat this exercise three times anti-clockwise and three times clockwise.

You should now feel considerably more relaxed. You will notice that you have been working up from the toes to the head and neck. Now you are ready to commence with stage two.

STAGE TWO *(time approx. 3 mins)*

1 Sit up straight in the chair, and take your pulse if necessary. If it is racing slightly, then allow yourself to relax until you feel more rested.

2   At this point in the relaxation programme, it is time to focus your thoughts in preparation for stage three. If you find that thought control with closed eyes is difficult, then open them and look at something that you admire which gives you pleasure. It might be some flowers, a photo of someone you love, or a view of the country through your window. If you want to keep your eyes closed, then start thinking about something helpful. It might be a happy song that you have heard on the radio during the last week, or a single word like 'love' or 'peace'. Perhaps you might find a phrase like 'I am feeling very relaxed and content' more effective. The aim is to bring your mind into a more peaceful condition and so help the body to relax even further.

3   Now exhale all the air in your lungs.

4   With your thoughts still focused on something pleasurable, perform ten complete breaths. You have ample time as the 4, 2, 4 ratio takes about $1\frac{1}{2}$ minutes, depending upon heart rate.

At this point you should feel much more relaxed. Your mind has been dwelling on images or ideas that are soothing your nerves, and your body has been calmed through stretching exercises and deep rhythmic breathing. Now for the more advanced visualization and suggestion techniques of stage three.

STAGE THREE *(time approx. 4 mins)*

If you can, continue to use the complete breath throughout this final stage. Should you find it difficult to concentrate on these mental exercises as well as the breathing, then breathe freely. If this is the case, when you become more proficient at these techniques, you might like to try and return to the complete breath – you can only benefit from it.

Because we are mentally diverse, this stage has been treated slightly differently. You will find a list of techniques that you can employ, some of them will suit you more than others. Use the ones that help you the most. Alternatively, invent some mental exercises of your own that follow along similar principles. You have the advantage of being able to make them directly applicable to your own life-style.

1   I am breathing harmonious forces right into my body that are continually around me. With each successive exhalation I am ridding myself of all tension and discord.

2   If I have any negative emotions towards anyone, I release them from my mind right now. I am at peace with all my friends and neighbours. Through my thoughts and actions I am making others more relaxed, fulfilled, and happy. I am, therefore, at the centre of happiness.

3   My whole body is feather-light and glowing with warmth. My heart and breathing rates have slowed right down. I am totally relaxed because I can feel all tensions leave my body with each positive thought.

4   My body is transparent and made of glass. Within me I have harboured negative emotions and attitudes that I now see as black ugly stains within my being. In my hands and feet I perceive valves that are beginning to open. After a while, my body is emptied of dark fluid and all the stains disappear. With each breath of fresh air I am filling my empty and clean glass body with brilliant light that leaves no area of me without illumination.

5   I am lying down in thick green grass in the countryside. There are open skies above me and a summer sun. Over my skin I feel the warmth of the sun's rays and the gentle afternoon breeze. The grass is a bed of peace and tranquillity to me. I am becoming a part of the nature that surrounds me and I am drinking in all the positive healing forces of the universe that envelop me at all times.

*   *   *

Within ten minutes you will be able to relax your mind and body by applying this system. Through regular use you will undoubtedly break-down long-term stress patterns that all of us unconsciously generate. Pay particular attention to critical days; if you can, find an extra few minutes to practise this during your working day, you will soon discover its calming yet vitalizing effect.

As you make positive steps to improve yourself through planning around highs, lows, and critical events, your life will become more efficient, safer, and fulfilling. There are hidden potentials within each of us, now you know how to nurture them to complete maturity. These pages guide you on the path, but you must make the first step. Begin today and live more fully with biorhythm, knowing that there are countless others who have proved its inestimable value in their own lives.

APPENDIX OF TABLES
AND CHARTS

TABLE OF LEAP YEARS FROM 1900 TO 2000
(INCLUSIVE)

1904	1924	1944	1964	1984
1908	1928	1948	1968	1988
1912	1932	1952	1972	1992
1916	1936	1956	1976	1996
1920	1940	1960	1980	2000

TABLE OF DAYS IN EACH MONTH

January	31	July	31
February	28	August	31
March	31	September	30
April	30	October	31
May	31	November	30
June	30	December	31

TABLE OF BIORHYTHM COMPATIBILITY
PERCENTAGES

DAYS BETWEEN EACH CYCLE	PHYSICAL	EMOTIONAL	INTELLECTUAL
0	100	100	100
1	91	93	94
2	83	86	88
3	74	79	82
4	65	71	76
5	57	64	70
6	48	57	64
7	39	50	58
8	30	43	52

DAYS BETWEEN EACH CYCLE	PHYSICAL	EMOTIONAL	INTELLECTUAL
9	22	36	45
10	13	29	39
11	4	21	33
12	4	14	27
13	13	7	21
14	22	0	15
15	30	7	9
16	39	14	3
17	48	21	3
18	57	29	9
19	65	36	15
20	74	43	21
21	83	50	27
22	91	57	33
23	100	64	39
24		71	45
25		79	52
26		86	58
27		93	64
28		100	70
29			76
30			82
31			88
32			94
33			100

PHYSICAL BIOGUIDE

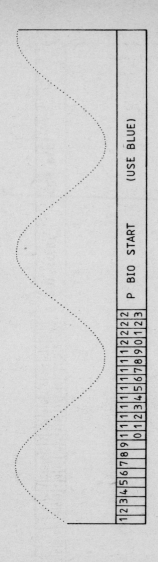

P BIO START   (USE BLUE)

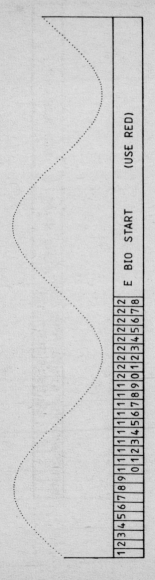

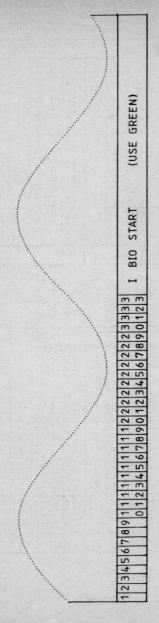

I BIO START   (USE GREEN)

1 2 3 4 5 6 7 8 9 1 1 1 1 1 1 1 1 1 1 2 2 2 2 2 2 2 2 2 2 3 3 3
0 1 2 3 4 5 6 7 8 9 0 1 2 3 4 5 6 7 8 9 0 1 2 3

119

BLANK BIOCHARTS

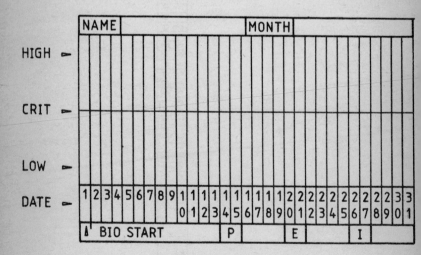

BLANK BIOCHART

The chart below is the *correct* size for use with the three Bioguides on pages 117-119 and in conjunction with the instructions on pages 42-46.

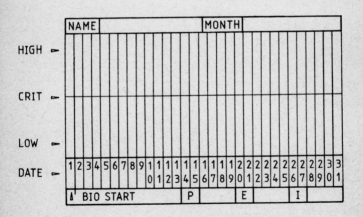

BLANK BIOCHART DAY COUNT

NAME		DOB	
		MONTH	
1	YEARS		=
2	YEARS		=
3	(1) OR (2) × 365		=
4	DAYS TO END MONTH		=
5	DAYS IN MONTHS		=
6	LEAP DAYS		=
7	TOTAL OF (3),(4),(5),(6) + 1		=
8	(7) / 23	(P)	=
	(7) / 28	(E)	=
	(7) / 33	(I)	=
9	(8P) × 23	(P)	=
	(8E) × 28	(E)	=
	(8I) × 33	(I)	=
10	(7) − (9P)	(P)	=
	(7) − (9E)	(E)	=
	(7) − (9I)	(I)	=

BLANK COMPATIBILITY DAY COUNT

PERS. 1		DOB		
PERS. 2		DOB		
1	YEARS × 365		=	
2	DAYS TO END MONTH		=	
3	DAYS IN MONTHS		=	
4	DAYS IN BIRTH MONTH		=	
5	LEAP DAYS		=	
6	TOTAL OF (1),(2),(3),(4),(5)		=	
7	(6) / 23	(P)	=	
	(6) / 28	(E)	=	
	(6) / 33	(I)	=	
8	(7P) × 23	(P)	=	
	(7E) × 28	(E)	=	
	(7I) × 33	(I)	=	
9	(6) − (8P)	(P)	=	
	(6) − (8E)	(E)	=	
	(6) − (8I)	(I)	=	
10	(9P)	(P)	=	%
	(9E)	(E)	=	%
	(9I)	(I)	=	%

BLANK BODY CHART

NAME		DATE	
1			
2			
3			
4			
5			
6			
7			
8			
9			
10			
11			
12			
13			
14			
15			
16			
17			
18			
19			
20			
21			
22			
23			
24			
25			
26			
27			
28			
29			
30			

GLOSSARY

BIORHYTHM   specifically, the physical, emotional and intellectual cycles.

BIOLOGICAL RHYTHM   biological cycles in general.

CIRCADIAN RHYTHM   a rhythm that has a frequency of about one day. From Latin *circa* = about, and *dies* = day, e.g. body temperature.

CIRCALUNAR RHYTHM   a rhythm that has a frequency which is similar to moon phasing, i.e. $29\frac{1}{2}$ days.

CIRCAMENSUAL RHYTHM – the menstrual rhythm that usually has a frequency between 28-30 days.

CIRCANNUAL RHYTHM   a rhythm that has a frequency of about a year, e.g. calcium excretion in Eskimos.

CRITICAL DAY   the period when a biorhythm changes phase from positive to negative or vice versa.

DISCHARGING PHASE   the period of a biorhythm when it is positive. Associated with higher energy states.

DOUBLE CRITICAL DAY   a period when any two of the three biorhythms change phase within the same 24 hours.

EMOTIONAL RHYTHM   the 28-day biorhythm.

FREE RUNNING   a situation where biological rhythms change their frequency due to a lack of external timing influence, e.g. changes in sleep/wake cycle in human isolation experiments.

HALF-PERIODIC DAY   sometimes used specifically to refer to the critical day that marks a phase change from positive to negative.

HOLISTIC   when used in alternative medicine the word refers to a healing approach that treats the entire person as an interactive whole, rather than the isolated ailment.

INTELLECTUAL RHYTHM   the 33-day biorhythm.

INTERNAL DESYNCHRONIZATION – a situation where various biological rhythms have become out of step with each other, e.g. someone jetting from USA to England will experience this as their body temperature/urine excretion/heart rhythms will be temporarily affected.

NEO-FLEISSIAN CYCLES   the rhythms attributed to Fleiss, but with modified interpretation, i.e. the physical, emotional, and intellectual cycles.

PHYSICAL RHYTHM the 23-day biorhythm.

RECHARGING PHASE the period of a biorhythm when it is negative. Associated with lower energy states.

SEMI-CAUTION DAY the days either side of a critical day, as these have been shown to present greater instability than non-critical days.

SENSITIVITY RHYTHM the 28-day biorhythm.

SUPRADIAN RHYTHM a rhythm that has a frequency which is greater than a day.

TRIPLE CRITICAL DAY a period when all three of the biorhythms change phase within the same 24 hours.

ULTRADIAN RHYTHM a rhythm that has a frequency of about 90 to 100 minutes, e.g. when sleeping, we alternate between REM and delta sleep every 90 minutes.

ZEITGEBER an environmental influence that is able to synchronize a biological cycle or behaviour pattern, e.g. light.

INDEX

Numbers in *italic* indicate illustrations.